U0309488

航天科工出版基金资助出版

常用标准件术语
中英对照词典

An English – Chinese Terminology
Dictionary of Standard Parts

航天精工股份有限公司　编著

王炎青　主编

李英亮　刘素云　马建军　主审

中国宇航出版社

·北京·

版权所有　侵权必究

图书在版编目（CIP）数据

常用标准件术语中英对照词典 ＝ An English－
Chinese Terminology Dictionary of Standard Parts /
航天精工股份有限公司编著；王炎青主编．--北京：
中国宇航出版社,2018.9

ISBN 978－7－5159－1530－2

Ⅰ.①常… Ⅱ.①航… ②王… Ⅲ.①航空器－标准
件－名词术语－对照词典－英、汉 Ⅳ.①V229－61

中国版本图书馆 CIP 数据核字（2018）第 215204 号

| 责任编辑 | 侯丽平 | 封面设计 | 宇星文化 |

出　版
发　行　　中国宇航出版社

社　址　北京市阜成路 8 号　　　　邮　编　100830
　　　　　（010）60286808　　　　　（010）68768548
网　址　www.caphbook.com
经　销　新华书店
发行部　（010）60286888　　　　　（010）68371900
　　　　　（010）60286887　　　　　（010）60286804（传真）
零售店　读者服务部
　　　　　（010）68371105
承　印　河北画中画印刷科技有限公司
版　次　2018 年 9 月第 1 版　　　2018 年 9 月第 1 次印刷
规　格　880×1230　　　　　　　开　本　1/32
印　张　9.375　　　　　　　　　字　数　270 千字
书　号　ISBN 978－7－5159－1530－2
定　价　68.00 元

本书如有印装质量问题，可与发行部联系调换

《常用标准件术语中英对照词典》
An English – Chinese Terminology Dictionary of Standard Parts
顾问委员会

主　任：赵　康

副主任：殷小健

顾　问：（按姓氏音序排列）

陈桂芳　韩碧勇　胡隆伟　胡庆宽

李立堂　马建宁　马　玲　谢丹贤

徐国奇

《常用标准件术语中英对照词典》
An English－Chinese Terminology Dictionary of Standard Parts
编写委员会

主　编：王炎青

主　审：李英亮　刘素云　马建军

委　员：（按姓氏音序排列）

程全士　樊金桃　焦　莎　刘敏丽

刘烨欣　卢　颖　宁广西　齐　跃

孙佳琪　王修保　徐　飞　徐　俭

徐英超　张晓斌　张晓玲

序

术语是表达或限定科学概念的约定性语言符号，是思想和认识交流的工具。本词典即是国内航空航天中高端工业基础件领军企业——航天精工股份有限公司策划的一本介绍标准件领域术语的专业图书。

近年来，航天精工坚守航空航天中高端工业基础件主业，以市场为导向，以科技创新为动力，以先进制造为支撑，工业化、信息化两化融合，专业化、规模化提质增效，牢牢占据国内市场龙头地位。同时，航天精工大力拓展国际化经营，积极响应"一带一路"倡议，与GE、庞巴迪等国际知名企业开展合作，对外交流日益增多，公司适时策划编写这本术语词典将有利于工程技术人员提升专业双语交流能力。

相较于一般术语词典，本词典更具实用性、专业性和针对性。首先，本词典认真总结了国内外标准件术语发展的成果，并结合了企业实践所积累的专业知识，提高了术语的准确性和适用性。其次，本词典涵盖了标准件设计、研发、制造等相关的标准化、机械加工、材料与热处理、表面处理、试验检测、质量控制等各个领域，提供了标准件行业相关的比较完备的专业术语。

希望本词典的出版，有益于标准件工程技术人员提升专业术语水平，有益于工程技术人员更好地引进、消化、吸收国外先进的标准件设计、制造技术，提升国内标准件水平。

航天精工股份有限公司董事长

2017 年 12 月

前　言

　　标准件是一种已经标准化并形成了标准的基础零部件，是装备制造业不可或缺的重要组成部分，直接影响装备和主机产品的性能水平、质量和可靠性，是《中国制造 2025》规定重点发展的"四基工程"之一——"工业基础件"的重要组成部分，是我国建设"制造强国"的重要基石和支撑。加快提升我国标准件研发生产水平，对我国未来五年、十年及以后的制造业发展具有十分重要的意义。而加快提升我国标准件研发生产水平，又必须充分利用国内外标准件发展的先进经验和科技成果。

　　自加入 WTO 以来，我国同世界各国的政治、经济、文化、技术交流日益增多。为了适应不断发展的国际交往需求，充分利用各种国际交流平台，积极引进、消化、利用国外先进的标准件设计、制造技术，实现与世界接轨，并以此为基础进行不断的改进和创新，持续提升我国标准件科研、生产、管理水平，更好地保证国家高新工程和装备发展要求，专业地使用标准件相关术语显得十分重要。为此，航天精工股份有限公司组织企业内外专家认真总结国内外标准件术语发展的成果，结合自身长期从事标准件科研生产经营所形成的专业知识编写了本词典。

　　本词典内容全面，涵盖产品与结构、标准化、机械加工、材料与热处理、表面处理、试验检测、质量管理等标准件设计、制造相关的各个专业领域。目前国内尚无涵盖内容如此系统的标准件专业术语词典，类似书籍多限于介绍标准件产品与结构。

　　本词典收录了常用标准件相关术语 3 000 余条。

　　本词典在结构上分为英-汉对照和汉-英对照两部分，英-汉对照部分，按英文字母顺序进行编排；汉-英对照部分，按产品与结构、标准化、机械加工、试验检测、材料与热处理、表面处理、质量管理、轴承等类别进行分类编排，每一类内部按汉语拼音顺序排列。同时，为保持各类术语相对完整，本词典刻意安排了少量术语在不同类别中重复出现。这样的安排使得本词典兼具英汉、汉英互查功能，且特别方便不同专业的使用人员进行快速检索。

本词典编写者所在企业——航天精工股份有限公司是中国航空航天中高端工业基础件领军企业，编写人员均来自科研生产一线，而且英语水平较高、工作经验丰富，并非常善于总结积累。词典融合了标准件相关术语标准发展的现有成果和编写者近十年的积累，并结合实际进行了甄别和选取，提高了术语的准确性和适用性。

本词典适合从事标准件科研、生产、经营、管理和使用的工程技术人员、管理人员和翻译人员使用，也适合大专院校相关专业的师生参考。

由于编者水平有限，错漏之处在所难免，恳请读者批评指正，以便我们再版时纠正。

编　者

2017 年 12 月

目 录

英文→中文

A 3

B 14

C 22

D 39

E 47

F 54

G 62

H 65

I 71

J 78

K 79

L 80

M 84

N 91

O 95

P 98

Q 108

R ·· 110

S ·· 117

T ·· 131

U ·· 139

V ·· 141

W ·· 143

X ·· 146

Y ·· 147

Z ·· 148

中文→英文

产品与结构 Products&Features ··················· 151

标准化 Standardization ·································· 167

机械加工 Machining ··································· 175

试验检测 Testing&Inspection ··················· 196

材料与热处理 Material&Heat Treatment ····· 229

表面处理 Surface Treatment ······················· 259

质量管理 Quality Management ···················· 273

轴承 Bearing ··· 279

英文→中文

A

abnormal grain growth	异常晶粒长大
above‐critical state	超临界状态
abrasion	磨耗/磨损
abrasion loss	磨耗量
abrasion resistance	耐磨性
abrasion resistance index	耐磨指数
abrasion resistant steel	耐磨钢
abrasive	磨料
abrasive blast cleaning	喷砂清理
abrasive tools	磨削刀具
absolute humidity	绝对湿度
absolute measurement	绝对测量
absolute pressure	绝对压力
absolute temperature	绝对温度
absolute value	绝对值
absolute zero	绝对零度（−273℃）
absorbed dose	吸收剂量
absorbed energy	吸收能量

absorption	吸收
accelerated ageing	加速时效/老化
accelerated combustion	加速燃烧
accelerated cooling	加速冷却
accelerated corrosion	加速腐蚀
accelerated creep	加速蠕变
accelerated cure	加速硫化
accelerated weathering test	加速天候老化试验
accelerator	促进剂
acceptable quality level(AQL)	验收/合格质量标准
acceptance	验收
acceptance inspection	验收检验
accreditation	认可
accumulated heat	积蓄热
acicular martensite	针状马氏体
acorn nut	盖形螺母
acquisition frequency	采集频率
acquisition time	采集时间
across corners	对角宽度
across flats	对边宽度
activated carbon	活性碳
activating treatment	活化处理

activation	激活/活化
activation polarization	电化学极化/活化极化
activator	活化剂
active atmosphere	活性气氛
active force	作用力
active thermal – graphic testing	主动式热像检测
activity	活度/活动
actual size	实际尺寸
addition polymerization	加聚反应
additive	添加剂
additive manufacturing（AM）	增材制造
adhesion	粘合/粘着
adhesion factor	粘合系数
adhesion of coating	涂层附着力
adhesion promoter	粘合促进剂
adhesion strength	粘合强度
adhesive	粘接剂
adjusting	调整/调节
administrative standard	管理标准
administrative standard system	管理标准体系
adoption	采用
adoption of a normative document	规范性文件的采用

advance product quality planning（APQP）	产品质量先期策划
aeolotropism	各向异性
Aerospace Industries Association（AIA）	（美）航空工业协会
Aerospace Material Specification（AMS）	（美）航空材料规范
Aerospace Standard（AS）	（美）航空标准
affinage furnace	精炼炉
affinity	亲和力
aftercure	后硫化/残余硫化
age hardening precipitation	时效硬化沉淀
age hardening steel	时效硬化钢
ageing	时效/老化
ageing stability	时效稳定性
agile manufacturing	敏捷制造
air cooling	空冷
air excess coefficient	空气过剩系数
air oven ageing	箱式热空气老化
air – hardening steel	空气淬硬钢
air quench	气淬
airscrew bearing	螺旋桨轴承
airworthiness	适航
aligning	找正
alkali bath furnace	碱浴炉

alkali metal	碱金属
alkali proof steel	耐碱钢
alkaline cleaning	碱洗
alloy ageing	合金时效
alloy segregation	合金成分偏析
alloy steel	合金钢
alloy tool steel	合金工具钢
aluminium（Al）	铝
ambient temperature	环境温度
amendment	修正
American Iron and Steel Institute（AISI）	美国钢铁学会
American National Standard Institute（ANSI）	美国国家标准学会
American Society for Testing and Materials（ASTM）	美国材料与试验协会
American Society of Mechanical Engineers（ASME）	美国机械工程师协会
American Society of Metals（ASM）	美国金属学会
amount of cold plastic deformation	冷塑性变形量
amount of cold work	冷加工量
amount of contraction	收缩量
amount of deflection	挠度
amount of deformation	变形量

amount of porosity	孔隙度
amount of vacuum	真空度
amplitude	振幅
analysis of spectra	光谱分析
analysis of technological efficiency	工艺性分析
anchor bolt	锚栓
angle of tilt	偏转角度
angular bearing	角接触轴承
angular contact ball bearing	角接触球轴承
angular contact spherical roller bearing	角接触球面滚子轴承
angular contact thrust ball bearing	角接触推力球轴承
angular contact thrust spherical plain bearing	角接触推力关节轴承
anisotropic material	各向异性材料
anisotropy	各向异性
annealing	退火
annealing brittleness	退火脆性
annealing cycle	退火周期/制度
annealing furnace	退火炉
annealing operation	退火工序
annealing - pickling line	退火酸洗线
anode	阳极
anodic polarization	阳极极化

anodic protection	阳极保护
anodizing	阳极(氧)化
anti – blocking agent	抗粘连剂
anti – coagulant	抗凝剂
antioxidant	抗氧剂
antistatic agent	抗静电剂
anvil	铁砧/砧座
apparent density	表观密度
apparent hardness	表观硬度
apparent porosity	表面孔隙率
appearance	外观
application of a normative document	规范性文件的应用
applied skin	覆皮
approach	切入量
approval	批准
aqueous wash	水洗
arbor	刀杆
arc welding	电弧焊
arch	炉顶
area of coverage	覆盖区
area of indentation	压痕面积
area of interest	评定区

Argon – Oxygen Decarburization（AOD）	氩氧脱碳
arm	摇臂
aromatic oil	芳烃油
arrangement	排布
artifact	假缺陷
artificial ageing	人工时效
artificial defect	人工缺陷
artificial drying	人工干燥
artificial sea water	人造海水
artificial weathering	人工气候老化
as – cast metal	铸态金属
ash	灰分
asphalt	沥青
asphalt rubber	沥青橡胶
as – quenched hardness	淬火硬度
as – received condition	交货状态
as – reduced metal	还原态金属
as – rolled condition	轧制状态
as – rolled end	辗制末端
assembly	装配(件)/组件
assembly datum	装配基准
assembly flow charts	装配流程图

assembly process	装配过程
assembly torque	安装力矩
association	团体/协会
association standard	团体标准
assurance standard	保障标准
atmospheric corrosion	大气腐蚀
atmospheric exposure test	大气暴露试验
atmospheric pressure	大气压
atomic absorption spectrometry（AAS）	原子吸收光谱法
atomic emission spectrometry（AES）	原子发射光谱法
atomic symbols	元素符号
atomizing	雾化
attemperator	恒温箱
audit	审核
audit client	审核委托方
audit conclusion	审核结论
audit criteria	审核准则
audit evidence	审核证据
audit finding	审核发现
audit plan	审核计划
audit programme	审核方案
audit scope	审核范围

audit team	审核组
auditee	受审核方
auditor	审核员
austenite	奥氏体
austenite stabilization	奥氏体稳定化处理
austenitic Ni – Cr stainless steel	奥氏体镍铬不锈钢
austenitizing	奥氏体化
autoclave	硫化罐
automated machining cell	自动化加工单元
automated production	自动化生产
automated storage/retrieval system (AS/RS)	自动化仓库
automatic feed	自动进给
automatic spray machine	自动喷涂机
automation	自动化
auxiliary electrode	辅助电极
auxiliary step	辅助工步
average grading	平均粒度
average particle diameter	平均粒径
average temperature difference	平均温差
average tempering	中温回火
axial clearance	轴向游隙
axial constants	晶轴常数

axial deformation	轴向变形
axial displacement	轴向位移
axial distance between the two center lines of the innerring grooves	沟心距
axial load	轴向载荷
axial permissible static load	轴向额定静载荷
axial preload	轴向预载荷
axial strain	轴向应变
axial stress	轴向应力
axial tensile	轴向拉伸
axial ultimate static load	轴向极限静载
axiality	同轴度
axis of pitch diameter	中径轴线
axis of thread	螺纹轴线
axisymmetric deformation	轴对称变形
axisymmetric drawing	轴对称拉伸
axisymmetric extrusion	轴对称挤压
axle neck	轴颈
axle shoulder	轴肩

B

back stress	反向应力
bainite	贝氏体
baking	烘干
balance of heat	热平衡
balanced steel	半镇静钢/半脱氧钢
ball bearing	球轴承
ball mill	球磨机
ball peening	喷丸硬化
ball socket bearing	耳轴轴承
band theory	能带理论
banded structure	带状组织
bands of deformation	形变带
bank	堆积胶
bar	棒材/棒料
bar code	条形码
bareness	缺胶
barrel burnishing	滚光
barrel plating	滚镀

barrel polishing	滚筒抛光
base	底座
base metal	母材金属
base oil	基油
basic hole	基准孔
basic profile	基本牙型
basic shaft	基准轴
basic size	基本尺寸
basic standard	基础标准
basis material	基体材料
basis metal	基体金属
batch	批次
batch annealing	分批退火
batch number	批号
batch type furnace	间歇式炉
bath furnace	浴炉
bearing	轴承
bearing cage	轴承保持架
bearing clearance/play	轴承游隙
bearing fatigue point	轴承疲劳极限
bearing flaking	轴承滚道表面剥落
bearing flange	轴承凸缘

bearing for aircraft engine	航空发动机轴承
bearing for aircraft control system	飞机操作系统轴承
bearing for aircraft landing gear	飞机起落架轴承
bearing in pairs	成对双联轴承
bearing lining/liner	轴承衬套
bearing lock washer	轴承锁紧垫圈
bearing locknut	轴承锁紧螺母
bearing over loading	轴承过载
bearing performance	轴承性能
bearing power	轴承承载能力
bearing quick release latch	承力快卸锁
bearing ring	套圈
bearing seizure	轴承咬死
bearing steel	轴承钢
bearing stress	支承应力
bearing surface/face	支承面
bearing with cylindrical bore	圆柱孔轴承
bearing with extended inner ring	宽内圈轴承
bearing with tapered bore	圆锥孔轴承
bell furnace	罩式炉
bench marks	标记/标线
bench work	钳加工

bench – work tool	钳工工具
bend radius	弯曲半径
bend test	弯曲试验
bending	弯曲
bending creep test	弯曲蠕变试验
best – size thread wire or ball	最佳螺纹量针或量球
bihexagon	十二角
bihexagonal head	十二角头
bilinear bolt	双线螺栓
bimetallic rivet	双金属铆钉
bimetallic article	双金属制品
binary phase diagram	二元相图
Bureau International des Poids et Mesures (BIPM)	国际计量局
black dot	黑斑/黑点
black oxide finish	氧化发黑处理
black scorch	炭黑焦烧
blank	毛坯/胶坯
blank drawing	毛坯图
blanking	冲裁/落料
blanking force	冲裁力
bleed out	渗出

blend radius	转接半径
blind bolt	盲螺栓
blind hole	盲孔
blind rivet	抽芯铆钉/盲铆钉
blind rivet with break mandrel	带截断心轴的抽芯铆钉
blind side clearance	单面间隙
blister	气泡
blocking	粘连
blocking/masking medium	遮挡介质
bloom	喷霜
blow molding	吹塑
blowing agent	发泡剂
blue annealing	发蓝退火
body	钉体
body centred cubic (BCC)	体心立方(晶格)
boiling steel	沸腾钢
bolt	螺栓
bond integrity of the liner	衬垫粘接完善性
bonding agent	粘接剂
boriding	硼化
boring	镗孔/镗削

boring machine/borer	镗床
boron (B)	硼
bottom view	仰视图
bound rubber	结合橡胶
brass	黄铜
break line	波浪线
breakaway torque	拧出/松脱力矩
breaking sharp corners	倒钝锐边
break - off	折断
bridge type baking oven	桥式烘干室
bright annealing	光亮退火
bright dipping	浸亮
bright heat treatment	光亮热处理
bright plating	光亮电镀
bright quenching/hardening	光亮淬火
brightener	抛光剂/增亮剂
Brinell Hardness (HB)	布氏硬度
British Iron and Steel Institute (BISI)	英国钢铁协会
British Standards Institution (BSI)	英国标准学会
brittle erosion behaviour	脆性腐蚀特性
brittle fracture initiation	脆性断裂源
brittle fracture/rupture	脆性断裂

brittle – ductile transition temperature	脆性–塑性转变温度
brittleness/brittlement	脆性/脆化
brittleness temperature	脆性温度
broaching machine/broacher	拉床
broaching/pull broaching	拉削
broken hardening	分级淬火
bronze	青铜
Brookfield viscometer	布氏（B 型）粘度计
brush application/coating	刷涂
brush plating	刷镀
bubble	气泡
buffings	橡胶屑
build up jigs/fixtures	组合夹具
build – up factor	积累因子
bulbed blind rivet	鼓包型抽芯铆钉
bulk coating	散装涂镀
bulk density	堆积密度
bulk heat treatment	整体热处理
bulk modulus	体积弹性模量
bumping	放气
burner	烧嘴
burning	过烧

burr	毛刺
bursting strength	爆破强度
bush	衬套
button head	平圆头
by – product	副产品

C

cadmium（Cd）	镉
calender	压延机
calendering	压延
calibrate	校准/标定
calibration block	校准试块
caliper	卡尺
capability	能力
carbide	碳化物/硬质合金
carbon（C）	碳
carbon black	炭黑
carbon constructional quality steel	优质碳素结构钢
carbon content	碳含量
carbon fibre	碳纤维
carbon profile	碳含量分布
carbon tool steel	碳素工具钢
carbon - free ferrite	无碳铁素体
carbon - free stainless steel	无碳不锈钢
carbon - free surface layer	表面脱碳层

carburized bearing steel	渗碳轴承钢
carburized case	表面渗碳层
carburizing	渗碳
carried fluid	载液
carrier gas	载气
case hardening	表面硬化
case–hardened bearing steel	表面硬化轴承钢
casting	铸件/铸造/浇注
casting alloy	铸造合金
cathode	阴极
cathodic electro deposition	阴极电泳涂装
cathodic polarization	阴极极化
cathodic protection	阴极保护
cathodic reaction	阴极反应
cation	阳离子
caustic embrittlement	碱脆
cavity	模腔/型腔
cell	孔眼/晶胞
cellular polymeric flexible material	多孔柔性聚合材料
centreless external cylindrical grinding machine	无心外圆磨床
centreless grinding (CG)	无心磨

centrifuge	离心干燥机
ceramic - matrix composites（CMCs）	陶瓷基复合材料
cetyl alcohol	十六醇
chain conveyor furnace	链条输送式炉
chalking	粉化
chamfer	倒角
chamfered end	倒角端
change control	更改控制
change order for technological documentation	工艺文件更改通知单
characteristic	特性
characteristic cooling curve	特性冷却曲线
characteristic factor	特性因素
charge	炉料
charge transfer system	炉料转移系统
charging basket	料筐
charging tray	料盘
Charpy notch impact test	夏比缺口冲击试验
Charpy pendulum impact test	夏比摆锤冲击试验
check crack	收缩裂纹
check for zero	零点校验
check gauge	校对规
cheese head	低圆柱头

chemical absorption	化学吸附
chemical attack	化学侵蚀
chemical bonding	化学粘接
chemical brightening	化学抛光
chemical cleaning	化学清洗
chemical conversion coating	化学转化膜
chemical corrosion	化学腐蚀
chemical oxidation	化学氧化
chemical passivation	化学钝化
chemical plating	化学镀
chemical polishing	化学抛光
chemical vapor deposition (CVD)	化学气相沉积
chip	残屑
chip conveyor	排屑装置
chisel	錾子
chlorinated polyethylene	聚氯乙烯
chloroprene rubber	氯丁橡胶
chromate anodic oxidation	铬酸阳极化
chromate primer	铬酸盐底层涂料
chromating	铬酸盐处理/铬化
chromatographic analysis	色谱分析
chromium (Cr)	铬

chromium depletion	贫铬
chuck	卡盘
circulation cooling system	循环冷却系统
circumferential field	周向磁场
circumferential groove	定位槽/环槽
clad aluminum	包铝
clamp	卡箍
clamping force	夹紧力
classification standard	分类标准
cleaning	清洗
cleaning line	清洗作业线
clearance	间隙
clearance fit	间隙配合
clevis pin	销轴
clip bolt	卡箍螺栓
close tolerance	紧公差
closed – end blind rivet	封闭型抽芯铆钉
cloth mark	布纹
coagulant	凝固剂
coagulant dipping	凝固剂浸渍
coagulation	凝聚
coarse pitch thread	粗牙螺纹

coarse - grained steel	粗晶粒钢
coarsening	粗化
coated fabric	涂覆织物
coating	涂覆/涂层
coating process	涂覆工艺
coating tool	涂层刀具
cobalt (Co)	钴
code of practice	规程
code of practice standard	规程标准
cognizant activity	认定机构
cohesion of coating	涂层结合力
coil	卷料
coil turns	线圈匝数
coiling test	卷曲试验
cold ageing	冷时效/常温时效
cold brittleness	冷脆性/低温脆性
cold coiling	冷卷
cold creep	低温蠕变
cold cure	冷硫化
cold curing	室温固化
cold deformation	冷变形
cold deformation strengthening	冷变形强化

cold drawing	冷拔/冷拉
cold extrusion	冷挤压
cold finish	冷精整
cold flow	冷态流变
cold forging	冷锻
cold formability	冷成形性
cold heading/upsetting	冷镦
cold rolling	冷轧
cold shut	冷隔
cold straightening	冷矫直
cold wall vacuum resistance furnace	冷壁/内热式真空电阻炉
cold work	冷加工/冷作
cold work hardening	冷加工硬化
cold – rolled steel sheet	冷轧钢板
collar	高锁螺母/凸缘
collet chuck	夹头
colorimetry	比色法
colour difference	色差
colour fastness	色牢度
colour index	颜色/比色指数
colourant	着色剂

colour – fastness on exposure to light	暴光色牢度
colouring	着色
combined bearing	组合轴承
combined machining(CM)	复合加工
combined stress	复合应力
combined sulfur	结合硫
combustion chamber	燃烧室
commercial metal	工业金属
commercially pure titanium	工业纯钛
company standard	企业标准
comparable standards	可比标准
comparative measurement	比较测量
compatibility	兼容性
compensating	补偿
compensation coil	补偿线圈
complete thread	完整螺纹
complex compound	络合物
complex high – speed steel	多元高速钢
complex shear modulus	复数剪切模量
compliance	柔量
component analysis	分量分析
composite material	复合材料

composite plating	复合电镀
compound	混炼胶
compound machining	复合加工
compounding	混炼
compressed air drying	压缩空气干燥
compression modulus	压缩模量
compression moulding	模压
compression set	压缩永久变形
compression strain	压缩应变
compression stress	压缩应力
compression stress relaxation	压缩应力松弛
compressive modulus of elasticity	压缩弹性模量
compressive strength	抗压强度
computer integrated manufacturing system (CIMS)	计算机集成制造系统
computer numerical control(CNC)	计算机数控
computer - aided design (CAD)	计算机辅助设计
computer - aided engineering (CAE)	计算机辅助工程
computer - aided manufacturing (CAM)	计算机辅助制造
computer - aided process planning (CAPP)	计算机辅助工艺设计
computer - aided test (CAT)	计算机辅助测试
concentrates	浓缩物

concentration	浓度
concentration polarization	浓差极化
concentricity	同轴度
concession	让步
concurrent deformation	伴生变形
condensing collector	冷阱
conditioning	调节/调整
conditioning agent	调节剂
conductivity	导电性/电导率
cone point	锥端
confidence interval	置信区间
configuration	技术状态
configuration authority	技术状态管理机构
configuration baseline	技术状态基线
configuration management	技术状态管理
configuration object	技术状态项
conforming	合格品
conformity	合格
conformity certificate	合格证
consensus	协商一致
consistency	粘稠度
consistency of composition	成分一致性

constant	常数
constant pressure，volume and temperature (p. v. t.)	恒压、恒容和恒温
constant load SCC test	恒载荷应力腐蚀破裂试验
constant strain SCC test	恒应变应力腐蚀破裂试验
contact angle	接触角
contact corrosion	接触腐蚀
contact stain	接触污染
contaminant	污染物
continual improvement	持续改进
continuous annealing	连续退火
continuous cooling curve	连续冷却曲线
continuous drawing	连续拉拔
continuous furnace	连续式炉
continuous galvanizing line（CGL）	连续镀锌作业线
continuous mixer	连续混炼机
continuous processing line（CPL）	连续加工处理作业线
continuous quench and temper line	连续淬火回火作业线
continuous vulcanization	连续硫化
continuous vulcanizer	连续硫化机

contour forging	模锻/模压
contraction	收缩
contraction cavity	缩孔
contrast	对比度
contrast sensitivity	对比灵敏度
controlled atmosphere heat treatment	可控气氛热处理
conventional forming technique	常规成形工艺
conversion coating	转化膜
converting furnace	吹炼炉
conveyor belt furnace	传送带式炉
cooling chamber	冷却室
cooling schedule	冷却制度
cooperation part	外协件
coordinate grid method	坐标网格法
copper (Cu)	铜
core drilling	钻孔
cored bar	芯棒
cored cellular material	空心多孔材料
corner	刀尖
correction	勘误/纠正
corrective action	纠正措施
correlation coefficient	相关系数

corrosion	腐蚀
corrosion and heat resistant steel	耐蚀耐热钢
corrosion cell	腐蚀电池
corrosion current	腐蚀电流
corrosion depth	腐蚀深度
corrosion inhibitor	防腐蚀剂
corrosion potential	腐蚀电位/电势
corrosion protection	防蚀
corrosion resistance	耐蚀性
cotter pin	开口销
coulomb efficiency	库仑效率
coulombmeter	库仑计
counterboring	锪孔/锪削/扩孔
countercurrent rinsing	逆流漂洗
countersink	沉孔
countersinking	锪(沉)孔
countersunk head	沉头
couplant	耦合剂
coupler/coupling	(管)接头
coupling	耦合
coupling factor	耦合系数
coupling losses	耦合损失

cover latch	口盖锁
cover screw	面板螺钉
crack	裂纹
crack extension	裂纹扩展量
crack growth	裂纹增量
crack length	裂纹长度
cracking sensitivity composition	裂纹敏感性组成（成分）
crater	浅坑
crazing	龟裂/微裂
creaming	膏化
creep	蠕变
creep curve	蠕变曲线
creep elongation time	蠕变伸长时间
creep index	蠕变指数
creep rupture time	蠕变断裂时间
creep strength	蠕变强度
creep test	蠕变试验
crest	牙顶
crevice corrosion	缝隙腐蚀
critical constant	临界常数
critical current density	临界电流密度

critical deformation	临界变形程度
critical plane	临界面
critical rate of hardening	临界淬火速率
critical shear strain	临界切应变
critical strain	临界应变
critical stress	临界应力
critical defect	致命缺陷
critical part	关键零件
cropping	下料
cross hole	十字孔
cross rail	横梁
cross recessed pan head screw	十字槽盘头螺钉
cross section	横截面
cross/cruciform recess	十字槽
crucible	坩埚
cryo – grinding	低温研磨
crystallization	结晶
cubic system	立方晶系
cup head	扁圆头
cup point	凹端
cupping	冲压凹痕
cure	硫化/固化

cure rate	硫化速率
curemeter	硫化仪
Curie point	居里点
Curie temperature	居里温度
curing furnace	固化炉
current density	电流密度
current efficiency	电流效率
current magnetization method	电流磁化法
curvature radius	曲率半径
customer	顾客
customer satisfaction	顾客满意度
cutter relieving	让刀/抬刀
cutting	切削加工
cutting fluid	切削液
cutting force	切削力
cutting power	切削功率
cutting temperature	切削温度
cutting tool	刀具
cutting tooling	切削加工工装
cycle	循环
cylindrical head	圆柱头
cylindrical outside surface of an outer ring	外圈外圆柱表面

cylindrical roller	圆柱滚子
cylindrical roller bearing	圆柱滚子轴承
cylindrical thread	圆柱螺纹
cylindrical thrust roller bearing	推力圆柱滚子轴承
cylindricity	圆锥度/圆柱度

D

damping block	阻尼块
damping coefficient	阻尼/衰减系数
damping constant	阻尼常数
data base	数据库
dated reference (to standards)	(对标准的)注日期引用
datum	基准/数据
dead metal zone	死区
dead steel	全脱氧钢/全镇静钢
dead zone	盲区
deburring	去毛刺
decarburization	脱碳
decibel (dB)	分贝
decolorization	脱色
deep groove ball bearing	深沟球轴承
defect	缺陷
defect detection sensitivity	缺陷检出灵敏度
definition	清晰度

deflashing	修边
deformation	变形
deformation texture	变形织构
deformation twin	变形孪晶
degradation	降解
degrade	降级
degreasing	除油
degree of finish	光洁度
degree of ionization	电离度
degree of saturation	饱和度
degree of undercooling	过冷度
degrees of correspondence	一致性程度
deionized water	去离子水
delamination	脱层/剥离
delayed sweep	延时扫描
demagnetization	退磁
denitriding	退氮
densitometer	光学密度计
density	密度
dependability	可信性
depolarization	去极化
deposition rate	沉积速率

depth of hardening zone	淬硬层深度
descale	除鳞/去氧化皮
descaling/derusting	除锈
desiccant	干燥剂
design and development	设计和开发
design assignment	设计任务书
design datum	设计基准
design profile	设计牙型
design review	设计评审
destructive test	破坏性试验
determination	确定
developing time	显像时间
development	制定/显影
deviation	偏差
deviation from spherical form	球形偏差
deviation in pitch	螺距偏差
deviation permit	偏离许可
devulcanization	脱硫
dew point	露点
diameter run - out	径向跳动
diamond knurl	网纹滚花
diaphragm	隔膜

die	模具
die forging	模锻(件)
die releasing	脱膜
dielectric constant	介电常数
differential filter	微分滤波器
differential heating	差温加热
differential thermal analysis（DTA）	差热分析
diffusion	扩散
diffusion annealing	扩散退火
diffusion control	扩散控制
diffusion hardening	扩散硬化
diffusion layer	扩散层
diluent	稀释剂
dilution stability	稀释稳定性
dimensional chain	尺寸链
dimensional tolerance	尺寸公差
dimple	压窝
DIN（Deutsches Institut für Normung）	德国标准化学会
dip coating	浸涂
dipping	浸渍
direct application of an standard	标准的直接应用
direct contact method	直接接触法

direct current plasma (DCP)	直流等离子体
direct exposure imaging	直接曝光成像
direct metal laser sintering (DMLS)	直接金属激光烧结
direct moulding technology	直接成型技术
directivity	指向性
discontinuity	不连续(性)
discontinuous furnace	间歇式炉/非连续式炉
dishing	冲压凹痕
dislocation	位错
dispersed phase hardening	弥散相硬化
dispersing agent	分散剂
dispersion	色散
dispersion medium	色散介质
displacement characteristic value	位移特征值
display area	显示区域
display through	穿透显示
dissolution	溶解
dissolved oxygen	溶解氧
distilled water	蒸馏水
distributor	经销商
dividing	刻线

dividing movement	分度运动
dose equivalent	剂量当量
dosemeter	剂量器
double end stud	等长双头螺柱
double row angular contact ball bearing	双列角接触球轴承
double row bearing	双列轴承
double row radial ball bearing	双列向心球轴承
double row self‐aligning ball bearing	双列调心球轴承
double row tapered roller bearing	双列圆锥滚子轴承
double shear	双剪
double‐direction thrust ball bearing	双向推力球轴承
double‐frequency induction heating equipment	双频感应加热装置
double‐lug anchor nut	双耳托板螺母
double‐shear joint	双剪接头
down draft spray booth	下吸式喷漆室
draft standard	标准草案
drafting	起草
drain time	流滴时间
drawing	拉拔
drawing furnace	牵引式炉

drift – expanding test of tube	金属管扩口试验
drilled – out rivet	钻出铆钉
drilling	钻削
drilling machine/driller	钻床
drive pin	击入式钉芯
driveability	铆接性
driving feature	扳拧形状
drop bottom furnace	底开式炉
drop feed atmosphere	滴注式气氛
dropping corrosion test	点滴腐蚀试验
dry cutting	干式切削
dry developer	干显像剂
dry grinding	干磨
dry method	干法
dry spray booth	干式喷漆室
drying	干燥
drying oven	干燥箱
dual phase steel	双相钢
dual – liquid quenching tank	双液淬火槽
ductility	延性
duplex alloy	二相合金
duplex annealing	双重退火

duplex austenitic – ferritic stainless steel	双相奥氏体铁素体不锈钢
duplex quenching tank	双联淬火槽
durability	持久性/耐用性
duration	持续时间
durometer	硬度计
dust collector	除尘装置
duty standard	工作标准
duty standard system	工作标准体系
dye penetrant	着色渗透剂
dynamic analysis	动态分析
dynamic balance test	动平衡试验
dynamic currents	动态电流
dynamic fatigue	动态疲劳
dynamic load	动载荷
dynamic measurement	动态检测
dynamic strain	动态应变
dynamic vulcanization	动态硫化

E

earing test	凸耳试验
eccentric angle	偏心角
eccentricity	偏心
eddy current	涡流
eddy current testing	涡流检测
edge distance	边距
edge effect	边缘效应
editorial change	编辑性修改
effective depth of penetration	有效渗透深度
effective magnetic permeability	有效磁导率
effective thread	有效螺纹
effectiveness	有效性
efficiency	效率
effusion	渗透
ejector pin	推杆
elastic deformation	弹性变形
elastic limit	弹性极限
elastic medium	弹性介质

elastic shear modulus	弹性剪切模量
elastic strain	弹性应变
elasticity	弹性
Electric Arc Furnace（EAF）	电弧炉
electric double layer	双电层
electric furnace	电炉
electrical drainage protection	排电流保护
electrochemical corrosion	电化学腐蚀
electrochemical equivalent	电化当量
electro – chemical machining（ECM）	电解加工
electrochemical polishing	电化学抛光
electrochemical protection	电化学保护
electrochemical treatment	电化学处理
electrochemistry	电化学
electrode	电极
electro – discharge machining（EDM）	电火花加工
electrolyte	电解质
electrolytic cleaning	电解清洗
electrolytic degreasing	电解除油
electrolytic grinding	电解磨削
electrolytic machining	电解加工
electrolytic pickling	电解酸洗

electrolytic polishing	电解抛光
electrolytic solution	电解液
electrolyze	电解
electromagnet	电磁铁
electromagnetic agitation	电磁搅拌
electromagnetic coupling	电磁耦合
electromagnetic radiation	电磁辐射
electromagnetic testing	电磁检测
electro–mechanical coupling factor	机电耦合系数
Electron Beam Furnace（EBF）	电子束炉
electron beam machining（EBM）	电子束加工
electrophoretic deposition	电泳沉积
electroplating	电镀
electropolishing	电（解）抛光
Electroslag Remelting（ESR）	电渣重熔
electrostatic spraying	静电喷涂
elemental analysis	元素分析
elements for aligning tool	对刀件
elevated temperature	高温
elevated temperature double shear test	高温双剪试验
elevated temperature tensile	高温拉伸
elliptical bearing	椭圆轴承

elongation	伸长(率)
elongation at a given stress	定应力伸长率
elongation at break/rupture	断裂伸长率
elongation at yield	屈服伸长率
emulsifiable penetrant	乳化渗透剂
emulsification	乳化
emulsification time	乳化时间
emulsifier	乳化剂
emulsion degreasing	乳化除油
emulsion polymerization	乳液聚合
EN (Europäische Norm)	欧洲标准
end	末端
end effect	端部效应
end face runout	端面跳动
endothermic atmosphere	吸热式气氛
engagement	积极参与
engineer test	工艺试验
engineering change order	工程更改单
engineering strain	工程应变
engineering stress	工程应力
enterprise standard system	企业标准体系
entry torque	拧入力矩

enveloping thread	包络螺纹
epoxide rubber	环氧橡胶
epoxy primer	环氧底漆
epoxy resin	环氧树脂
equiaxed structure	等轴组织
equilibrium electrode potential	平衡电极电势
equivalent	当量
equivalent method	当量法
Erichsen cupping test	埃里克森杯突试验
erosion	侵蚀
error	误差
etch	蚀刻/侵蚀/腐蚀
ethyl acetate	乙酸乙酯
ethylene propylene rubber（EPR）	乙丙橡胶
eutectic structure	共晶组织
eutectic‐peritectic reaction	共晶-包晶反应
eutectoid steel	共析钢
evaluation	评定
evaporation	蒸发/汽化
exclusive requirement	必达要求
exhaust air washing system	排气洗净装置
exothermic atmosphere	放热式气氛

expanded rivet	膨胀铆钉
expanding	扩孔
expansion	膨胀
expansion fitting	冷装
exposure	曝光/暴露
exposure chart	曝光曲线
exposure test	耐候老化试验/户外暴露试验
extended inner ring	宽内圈
extender	填充剂
extensibility	延展度
extension	延伸
extension bar	接杆
extensometer	伸长计/引伸计
external cylindrical grinding	外圆磨
external thread	外螺纹
extra large size bearing	超大型轴承
extra light series bearing	轻系列轴承
extra – fine steel	特优钢
extra – high tensile steel	超高强度钢
extruded shapes	挤压型材
extruding	挤压

| extrusion | 挤出 |
| eye bolt | 活节螺栓 |

F

face seal	端面密封
face – centered cubic (FCC)	面心立方(晶格)
failure	失效
failure analysis	失效分析
failure model and effect analysis (FMEA)	失效模式与影响分析
false boss	工艺凸台
family	族
fastener	紧固件
fatigue	疲劳
fatigue breakdown	疲劳破坏
fatigue deformability	疲劳变形能力
fatigue ductility coefficient	疲劳延性系数
fatigue life	疲劳寿命
fatigue limit	疲劳极限
fatigue strength	疲劳强度
fatigue strength at N cycles	N 次循环后的疲劳强度
fatigue stress	疲劳应力
fatigue test	疲劳试验

feasibility	可行性
feasible test	可行性试验
feed	进给量
feed speed	进给速度
female die	阴模
ferrite	铁素体
ferritic matrix	铁素体基体
ferritic stainless steel	铁素体不锈钢
ferrous metal	黑色金属
field of standardization	标准化领域
filled – core rivet	填芯铆钉
filler	填料
filleting	倒圆角
filling slot	装球缺口
filling slot ball bearing	装填球槽球轴承
film thickness gauge	膜厚测量仪
filter	滤光板/滤波器/过滤器
final gauge length (after fracture)	断后标距
final heat treatment	最终热处理
final product	成品
fine grained steel	细晶粒钢
fine pitch thread	细牙螺纹

finish	表面处理
finish turning	精车(削)
finish/top coat	面漆
finished goods	制成品
finishing	精加工
finishing cut	光整加工
finite element analysis (FEA)	有限元分析
finite element modeling	有限元建模
firming/solidified agent	固化剂
first article assurance (FAA)	首件确认
first article inspection (FAI)	首件鉴定
fission	裂变
fissure	裂纹
fit	配合
fit shank	加强杆
fitness for purpose	适用性
fitting	软管接头
fitting allowance	装配余量
fixed datum	定位基准
fixing	定影
fixture	夹具
flaking	剥落

flammability	可燃性
flange	法兰
flange test	卷边试验
flanged bearing	凸缘轴承
flank	牙侧
flaring	扩口
flaring test	扩口试验
flash	闪镀/飞边
flash off	凉干（流平）
flash point	闪点
flash welding	闪光焊
flat cold rolled strip	冷轧带材
flat die thread rolling	搓丝
flat point	平端
flatness	平面度
flattening test of tube	金属管压扁试验
flexibility	柔度
flexible die	柔性模
flexible manufacturing cells (FMC)	柔性制造单元
flexible manufacturing system (FMS)	柔性制造系统
flexible manufacturing technology (FMT)	柔性制造技术
flexometer	挠曲试验机

flexure	弯曲/挠曲
floating anchor/plate nut	游动托板螺母
flow chart/diagram	流程图
flow marks	流痕
flue	烟道
fluid resistance	耐液性
fluidized bed carburizing	液态床渗碳
fluorescence	荧光
fluorescent dry deposit penetrant	干沉积荧光渗透剂
fluorescent inspection	荧光探伤
fluorescent magnetic powder	荧光磁粉
fluorescent penetrant	荧光渗透剂
fluorescent screen	荧光屏
fluorocarbon rubber	氟烃橡胶
flush rivet	沉头铆钉
flushness	平齐度
flux penetration	磁通穿透深度
focal distance	焦距
fog	灰雾
fog density	灰雾密度
force	力
forced air cooling device	风冷装置

forgeability	可锻性
forging	锻件/锻造
forging and stamping	锻压
forging flow line	锻造流线
form	牙型
form of section	断面形状
form of thread	螺纹牙型
formability	成形性
forming	成形
forms of shank	杆部形状
formulation	配方
forward extrusion	正挤压
foundation bolt	地脚螺栓
foundry	铸造
four point contact ball bearing	四点接触球轴承
four row tapered roller bearing	四列圆锥滚子轴承
fractograph	断口金相照片
fracture	断裂/断口
fracture grain size	断口晶粒度
fracture toughness	断裂韧性/韧度
frame	机身
free ferrite	游离铁素体

free sulfur	游离硫
free – machining steel	易切削钢
frequency	频率
frequency constant	频率常数
frequency & severity (F/S)	频率和严重度
friction	摩擦
friction bearing	滑动轴承
friction coefficient	摩擦系数
friction welding	摩擦焊
front vestibule	前室
frontal plane	正面
Fsu	剪切强度极限
Ftu	拉伸强度极限
full annealing	完全退火
full indicator movement (FIM)	表针摆动范围
full thread	全螺纹
full type ball bearing	满装球轴承
fundamental deviation	基本偏差
fundamental triangle	基本三角形
furnace body	炉体
furnace casing	炉壳
furnace chamber	炉室

furnace cooling	炉冷
furnace door	炉门
furnace frame	炉架
furnace lid	炉盖
furnace temperature stability	炉温稳定度
furnace temperature uniformity	炉温均匀度
furnace wall	炉墙
fused deposition modeling（FDM）	熔融沉积成型
fusion	熔化/（核）聚变

G

gage pressure	表压力
galvanic cell	原电池
galvanic corrosion	电偶腐蚀
galvanizing	热镀锌
gang channel floating anchor self‐locking nut	成组游动托板自锁螺母
gang channel nut	槽型螺母
gas purification equipment	气体净化装置
gas quenching medium	气体淬火介质
gaseous corrosion	气体腐蚀
gasket	衬垫
gas‐quenching vacuum resistance furnace	气淬真空电阻炉
gauge block	量块
gauge diameter	基准直径
gauge length	基准长度/标距
gauge plane	基准平面
gear cutters	齿轮刀具
gearbox	变速箱

gel	凝胶
general accuracy machine tools	普通机床
general assembly	总装
general normative elements	规范性一般要素
general yield displacement	屈服位移
geometric effect	几何效应
geometric unsharpness	几何不清晰度
geometrical products specifications (GPS)	产品几何技术规范
germ nucleus	晶核中心
glass reinforced plastics (GRP)	玻璃钢
gloss level	光泽度
gluing	粘接
GO gauge	通规
gouging	刨削
grade	等级/牌号
graded hardening	分级淬火
grain	晶粒
grain boundary	晶界
grain flow	晶粒流线
grain size	晶粒度
grain size number	晶粒号
grain - refinement treatment	晶粒细化处理

graphite	石墨
graphite furnace	石墨炉
graphitizing treatment	石墨化退火
gravity feed furnace	重力输送式炉
gravity spray gun	重力式喷枪
grinding	磨削/磨光
grinding machine/grinder	磨床
grip length	夹层长度
grip range	夹层范围
groove radius	沟道半径
grooved pin	槽销
grooving	切槽
group technology（GT）	成组技术
guide plate	导板
guide standard	指南标准
guideline	指南
guiding element	导向件
γ ray	γ 射线
γ transformation	γ 相变

H

half life	半衰期
Hall coefficient	霍尔系数
Hall effect sensor	霍尔效应传感器
hammer head	T形头
hard alloy	硬质合金
hardenability	淬透性
hardened and tempered steel	调质钢
hardening	硬化/淬火
hardening capacity	淬硬性
hardness	硬度
Haring cell	哈林槽
harmonic analysis	谐波分析
harmonized/equivalent standards	协调标准
hatch cover	口盖
head locking hole diameter	头部保险孔直径
head shapes	头部形状
heading	镦锻/顶镦
hearth	炉底

hearth plate	炉底板
heat affected area	热影响区
heat build – up	生热/热积累
heat distortion temperature（HDT）	热变形温度
heat equivalent	热当量
heat resistance	耐热性
heat sensitizer	热敏剂
heat transfer coefficient（HTC）	传热系数
heat treatment	热处理
heat treatment furnace	热处理炉
heating chamber	加热室/炉膛
heating medium	加热介质
heating schedule	加热制度
heating up time	升温时间
heat – sensitive dipping	热敏浸渍
heavy metals（H. M.)	重金属
helix	螺旋线
hexagon	六角
hexagon castle nut	六角皇冠螺母
hexagon head	六角头
hexagon head bolt with collar	六角凸缘螺栓
hexagon head bolt with flange	六角法兰面螺栓

hexagon head screw plug	六角头螺塞
hexagon head with collar	六角凸缘头
hexagon head with flange	六角法兰面头
hexagon head with washer face	六角垫圈面头
hexagon socket	内六角
hexagon socket head cap screw	内六角圆柱头螺钉
hexagon thin nut	六角薄螺母
hexagonal bar	六角钢
hexalobular	六角花形
hexalobular head	六角花形头
high alloy steel	高合金钢
high carbon chromium bearing steel	高碳铬轴承钢
high carbon martensite	高碳马氏体
high density inclusion (HDI)	高密度夹杂
high energy beam heat treatment	高能束热处理
high energy X-rays	高能 X 射线
high frequency induction heating equipment	高频感应加热装置
high pressure water cutting	高压水切割
high quality structural steel	优质结构钢
high speed steel (HSS)	高速钢
high strength low alloy steels (HLS)	高强度低合金钢

high temperature oxidation (HTO)	高温氧化
high temperature tempering	高温回火
high – cycle fatigue test	高周疲劳试验
high – duty metal (HDM)	高强度金属
high – energy – rate forming (HERF)	高速高能成形
high – locking/hi – lock bolt	高锁螺栓
high – locking/hi – lock nut	高锁螺母
high – speed cutting	高速切削
hindering of dislocation	位错受阻
holding	保温
hole broaching	拉孔
hole drilling	钻孔
hole grinding	磨孔
hole honing	珩孔
hole lapping	研孔
hole milling	铣孔
hole push broaching	推孔
hole rolling	滚压孔
hole slotting	插孔
hole turning	车孔
hole – basis system	基孔制
hollow	中空

hollow forging	空锻
hollow – core rivet	空心铆钉
homogeneous deformation	均匀变形
homogeneous steel	均质钢
homophase	同相
honing	珩磨
horizontal limit	水平极限
horizontal linearity	水平线性
horizontal location	水平定位
horizontal plane	水平面
horizontal vulcanizer	卧式硫化机
hose	软管
hot air drying	热风干燥
hot bath hardening	热浴淬火
hot dip	热浸镀
hot forging	热锻
hot heading	热镦
hot melting	热熔
hot short cracking	热脆开裂
hot short material	热脆材料
hot shortness zone	热脆区
hot spraying	热喷涂

hot wall vacuum resistance furnace	热壁/外热式真空电阻炉
hot working	热加工
hot – dip galvanized coating	热浸镀锌层
housing	轴承座
Hull cell	霍尔槽
human factor	人为因素
humidity	湿度
hydraulic jigs/fixtures	液压夹具
hydraulic moulding	液压成型
hydrofluoric acid	氢氟酸
hydrogen（H）	氢
hydrogen depolarization	氢去极化
hydrogen embrittlement	氢脆
hydrogen relief treatment	脱氢处理
hydrophilic emulsifier	亲水性乳化剂
hydro – thermal sealing	水合封孔处理
hypereutectoid steel	过共析钢
hypoeutectoid steel	亚共析钢
hysteresis	磁滞
hysteresis loop	滞后回线

I

identical (IDT)	等同
identical standards	等同标准
identification	标识/标志/鉴定
idle stroke	空行程
image contrast	图像对比度
image enhancement	图像增强
image magnification	图像放大
image quality	图像质量
immersion plate	浸镀
immersion rinse	浸没清洗
immersion test	全浸试验
impact resistance	抗冲击性能
impact test	冲击试验
impact toughness test	冲击韧性试验
improvement	改进
impurity	杂质
inch thread	寸制螺纹
inching/jogging	点动

inclusion	夹杂物
inclusion rating	夹杂等级
incoming inspection	入厂检验
incomplete annealing	不完全退火
incomplete thread	不完整螺纹
increased shank	加强杆
indentation	压痕
indentation hardness	压痕硬度
indentation modulus	压痕模量
indenter	压头
indication	指示
indirect application of an standard	标准的间接应用
indirect exposure	间接曝光
indirect resistance heating	间接电阻加热
induced surface hardening	感应加热表面淬火
induction annealing	感应退火
induction coil	感应线圈
induction hardening	感应淬火
induction tempering	感应回火
induction time	诱导期
inductive sensor	感应传感器
inductively coupled plasma (ICP)	电感耦合等离子体

industry standard	行业标准
inert anode	惰性阳极
inert atmosphere	惰性气氛
inert filler	惰性填料
inert gas	惰性气体
inflexion point	拐点/回折点
information system	信息系统
informative elements	资料性要素
infrared absorption spectra	红外吸收光谱
infrared camera	热像仪
infra – red furnace	红外炉
infrared radiation	红外辐射
infrared ray	红外线
infrared testing	红外检测
infrared thermal imaging	红外热成像
infrared thermography	红外热像法
inhibitor	抑制剂
initial stress	初始应力
injection moulding	注塑
inner ring	内圈
inner ring width	内圈宽度
innovation	创新

inorganic salt solution quenching medium	无机盐水溶液淬火介质
input process quality control (IPQC)	过程质量控制
insert	螺套/嵌件
inside diameter (ID)	内径
inspection	检查/检验
inspection lot	检验批
installation	安装
instantaneous stress	瞬时应力
Institute of Electrical and Electronic Engineers (IEEE)	(美)电器和电子工程师学会
instruction	指示
insulated layer	绝缘层
insulating material	绝缘材料
integrated manufacturing system	集成制造系统
intent heat of phase change	相变潜热
interchangeability	互换性
intercrystalline corrosion	晶间腐蚀
intercrystalline segregation	晶间偏析
intercrystalline shrinkage crack	晶间收缩裂纹
interelectrode distance	极间距
interested party	相关方

interface	接口/界面
interface standard	接口标准
interface/boundary	界面
interfacial tension	界面张力
interference	干涉
interference fit	过盈/干涉配合
intergranular attack (IGA)	晶间腐蚀
intergranular carbide precipitation	晶间碳化沉积
intergranular corrosion	晶间腐蚀
intergranular oxidation (IGO)	晶间氧化
intermediate heat treatment	中间热处理
intermetallic compound	金属间化合物
internal cylindrical grinding	内圆磨
internal mixer	密炼机
internal stress	内应力
internal thread	内螺纹
internally - heated bath furnace	内热式浴炉
International Civil Aviation Organization (ICAO)	国际民航组织
International Electrotechnical Commission (IEC)	国际电工委员会
International Institute of Welding (IIW)	国际焊接学会

International Iron and Steel Institute (IISI)	国际钢铁学会
International Organization for Standardization (ISO)	国际标准化组织
international rubber hardness degrees (IRHD)	国际橡胶硬度
international standard	国际标准
international standardization	国际标准化
international standards organization	国际标准组织
International Telecommunication Union (ITU)	国际电信联盟
internationally harmonized standards	国际协调标准
Internet of Things (IoT)	物联网
interphase boundary	相界面
inter – process annealing	工序间退火
interstage annealing	中间退火
inventory control system	库存控制系统
involvement	参与
ion beam machining (IBM)	离子束加工
ion implantation	离子注入
ion plating	离子镀
ion vapor deposition (IVD)	离子气相沉积
ion – bombarding heat treatment vacuum furnace	真空离子轰击热处理炉

ion – carburizing vacuum furnace	真空离子渗碳炉
iron (Fe)	铁
isothermal forging	等温锻
isothermal hardening	等温淬火
isothermal normalizing	等温正火
isothermal transformation	等温转变
isotope	同位素
isotropy	各向同性
item	产品/项(目)

J

Japanese Industrial Standard (JIS)	日本工业标准
Japanese Industrial Standards Committee (JISC)	日本工业标准调查会
Japanese Institute of Metals (JIM)	日本金属学会
jaw	卡爪
jigs	夹具
job shop	作业车间
joint	接头
joint audit	联合审核
joint radiation convection oven	辐射对流烘干室
Jominy end quenching test	端淬试验

K

kinematic path/travel	行程
kinematic viscosity	运动粘度/动力粘度
Knoop hardness（HK）	努氏硬度
know – how	专门技术/诀窍
knuckles	结块
knurl	滚花
knurled thumb screw	滚花高头螺钉

L

lack of coverage	露底
laminar fracture	层状断裂/断口
laminated object manufacturing	分层实体制造
lap	折叠/搭接
lap joint	搭接(接头)
lapping	研磨
large footprint thread – type blind rivet	大底脚螺纹型抽芯铆钉
laser beam machining (LBM)	激光加工
laser cladding	激光溶覆
laser electroplating	激光电镀
laser glazing	激光釉化
laser heating	激光加热
lathe fixture	车床夹具
lattice defect	晶格/点阵缺陷
lattice dilatation	晶格膨胀
lattice heterogeneity	晶格异质
lattice/grate	晶格
law of similarity	相似定律

lay	纹理方向
leaching	沥滤
lead	导程
lead（Pb）	铅
lead angle	导程角/升角
lead of helix	螺旋线导程
lead thread	引导螺纹
leading flank	引导牙侧
leak testing	泄漏检测
least material limit（LML）	最小实体极限尺寸
ledeburite	莱氏体
Leeb hardness（HL）	里氏硬度
life test	寿命试验
left－hand thread	左旋螺纹
length of coverage	覆盖长度
level of standardization	标准化层次
light fastness	耐光牢度
lightening hole diameter	减轻孔直径
limit deviation	极限偏差
limits of size	极限尺寸
linear attenuation coefficient	线性衰减系数
linear strain	线性应变

liner	衬垫
link rod pin	联杆销
lipophilic emulsifier	亲油性乳化剂
liquid film developer	液膜显像剂
load flank	承载牙侧
load – displacement curve	载荷-位移曲线
loading	上料
local area network (LAN)	局域网
local corrosion	局部腐蚀
local heat treatment	局部热处理
localized/selective carburizing	局部渗碳
locating elements	定位零件
locating pin	定位销
location	定位
lock washer	锁紧垫圈
locking	锁紧/收口
locking element	锁紧零件
locking feature	锁紧特性
locking ring	锁环
long dog point	长圆柱端
longitudinal resolution	纵向分辨率
lot pilot	批量试制

low temperature tempering	低温回火
low – alloy steel	低合金钢
low – cycle fatigue test	低周疲劳试验
lower limit deviation（EI）	下偏差
lower yield strength	下屈服强度
low – temperature impact brittleness test	低温冲击脆性试验
lubricant	润滑剂
lubrication groove	润滑槽
lubrication hole	润滑孔
luminance	亮度
luster	光泽

M

machinability	可加工性
machine load rate	设备负荷率
machined surface	已加工表面
machining	机械加工
machining accuracy	加工精度
machining allowance	机械加工余量
machining center（MC）	加工中心
machining complex	工艺系统
machining error	加工误差
macro examination	宏观检查
macrostructure	宏观组织
magnesium（Mg）	镁
magnetic field distribution	磁场分布
magnetic coil	磁化线圈
magnetic field	磁场
magnetic field indicator	磁场指示器
magnetic field meter	磁场计
magnetic field strength	磁场强度

magnetic flux	磁通
magnetic flux density	磁通密度
magnetic particle	磁粉
magnetic particle indication	磁痕
magnetic particle inspection	磁粉探伤
magnetic permeability	磁导率
magnetizing	磁化
major defect	严重缺陷
major diameter	螺纹大径
male die	阳模
malleability	展性
mandrel	钉芯
mandrel break load	钉芯断裂载荷
manganese（Mn）	锰
man - hour quota	工时定额
manual riveting	手工铆接
manual testing	手工检测
manufacturing discipline	工艺纪律
manufacturing equipment	工艺设备
manufacturing lot	生产批
manufacturing lot number	生产批号
maraging steel	马氏体时效钢

marine corrosion	海洋腐蚀
marking	标记
Martens hardness（HM）	马氏硬度
martensite	马氏体
masking	遮蔽
masking tape	遮蔽胶带
masonry bolt	开叉地脚螺栓
mass attenuation coefficient	质量衰减系数
mass spectrometer（MS）	质谱仪
mass spectrometry	质谱分析法
master alloy	中间合金
master batch	母炼胶
material consumption quota in process	材料消耗工艺定额
material profile	实体牙型
material requirement planning（MRP）	物料需求计划
matrix	基体
matrix precipitation	基体沉淀/析出
maximum limits of size	最大极限尺寸
maximum locking torque	最大锁紧力矩
maximum material limit（MML）	最大实体极限尺寸
maximum material profile	最大实体牙型
measurement	测量

measurement management system	测量管理体系
measuring apparatus	计量装置
measuring datum	测量基准
measuring instruments	计量仪器
mechanical cleaning	机械清洗
mechanical conditioning	机械调节
mechanical plating	机械镀
mechanical polishing	机械抛光
mechanical property/behaviour	力学/机械性能
mechanical sanding	机械打磨
mechanical seal	机械密封
mechanical strain	机械应变
mechanical testing	力学试验
mechanization	机械化
medium frequency induction heating equipment	中频感应加热装置
medium temperature tempering	中温回火
melt mixing	熔融混炼
mercury（Hg）	汞
metal flow	金属流动
metal injection moulding（MIM）	金属注射成型
metal plastic working	金属塑性加工

metallographic analysis	金相分析
metallographic examination	金相检验
metallography	金相(学)
metal – matrix composites（MMCs）	金属基复合材料
metering device	计量装置
methyl ethyl ketone	丁酮
metric thread	米制螺纹
metrological characteristic	计量特性
metrological confirmation	计量确认
metrological function	计量职能
micro indentation harness	微压痕硬度
micro – etching	显微腐蚀
micromachining	微细加工
micrometer	千分尺
microshrinkage	微缩孔
microstructure	微观结构/显微组织
microvoid	微空洞/微孔
milling	铣削
milling cutters	铣刀
milling machine/miller	铣床
mineral acid	无机酸
miniature bearing	微型轴承

minimum limits of size	最小极限尺寸
minimum material profile	最小实体牙型
minor defect	轻度缺陷
minor diameter	螺纹小径
minor phase	再生相/次生相
mission	使命
MJ profile thread	MJ 螺纹
modified（MOD）	修改
modulus	模量
modulus of elasticity	弹性模量
molar mass	摩尔质量
molybdenum（Mo）	钼
molybdenum disulfide（MoS_2）	二硫化钼
molykote	二硫化钼润滑剂
monitoring	监视
Mooney viscosity	门尼粘度
mother stock	母料
mould	模具
moulding	模塑/成型
moulding shrinkage	模压收缩
multicomponent diffusion agent	多元共渗剂
multi - crystal	多晶的

multilayer plating	多层电镀
multiphase	多相(的)
multirow bearing	多列轴承
multi – start thread	多线螺纹
multi – station automatic former	多工位自动成型机
multi – working zone furnace	多工区炉
mushroom head	扁圆头
mushroom head nib bolt	扁圆头带榫螺栓

N

nano – processing	纳米加工
National Aerospace and Defense Contractors Accreditation Program（NADCAP）	（美）国家航空航天和国防合同方授信项目
National Aerospace Standard（NAS）	（美）国家航空标准
national standard	国家标准
national standardization	国家标准化
national standards body	国家标准机构
natural ageing	自然时效/老化
natural atmosphere	自然气氛
natural rubber	天然橡胶
near surface defect	近表面缺陷
necking	缩颈
needle roller	滚针
needle roller bearing	滚针轴承
needle roller thrust bearing	推力滚针轴承
negative electrode	负极
negative segregation	负/反偏析

neutral atmosphere	中性气氛
neutral salt	中性盐
neutral salt spray (NSS) test	中性盐雾试验
new edition	新版本
nick bend test	缺口弯曲试验
nickel (Ni)	镍
niobium/columbium (Nb)	铌
nitric acid	硝酸
nitriding	渗氮
nitriding atmosphere	渗氮气氛
nitrile rubber	丁腈橡胶
nitrocellulose lacquer	硝基漆
nitrogen (N)	氮
NO GO gauge	止规
noble metal	贵金属
no – load test	空载试验
nominal diameter	公称直径
nominal frequency	标称频率
nominal size	公称尺寸
nominal strain	公称应变
nominal stress	公称应力
nonconformance report (NCR)	不符合项报告

non‐conforming parts per million（ppm）	每百万件中的不合格件数
nonconformity	不合格
non‐destructive testing（NDT）	无损检测
nonequivalent（NEQ）	非等效
non‐ferrous metal	有色金属
non‐metallic inclusion	非金属夹杂
nonrelevant indication	假象
non‐traditional machining（NTM）	特种加工
normal force	法向力
normal shank	标准杆
normal stress	法向应力
normalizing	正火
normative document	规范性文件
normative elements	规范性要素
notch	缺口
nozzle	喷嘴
null balance method	零位平衡
number of strokes	往复次数
numerical control machine tool	数控机床
numerical control machining	数控加工
numerical control（NC）	数控

nut	螺母
nylon washer	尼龙圈

O

objective	目标
objective evidence	客观证据
observer	观察员
occupational safety and health act (OSHA)	职业安全与卫生条例
octagon	八角
octagonal head	八角头
Oersted (Oe)	奥斯特
offset	残余变形
offset cruciform recess	偏心十字槽
offset yield point	残余屈服点
offset yield stress	残余屈服应力
oil bath furnace	油浴炉
oil quenching	油淬
oil sealing	油封
open – end blind rivet with countersunk head	开口型沉头抽芯铆钉
operation	工序
operation allowance	工序余量
operation datum	工序基准

operation diagram	工序图
operation dimension	工序尺寸
operation sheet	工序卡片
optimum cure	正硫化/最佳硫化
optional elements	可选要素
optional requirement	可选要求
ordinate	纵坐标
organic acid	有机酸
organization	组织
original equipment manufacturer（OEM）	原始设备制造商
original gauge length	原始标距
oscillating	摆动
outer ring	外圈
outside diameter（OD）	外径
outsource	外部
oval head	半沉头
over development	显影过度
over emulsification	过乳化
overcure	过硫化
overhanging rail	悬臂
overheating	过热
overload recovery time	过载恢复时间

oversize shank	加大杆
oversize shank	加大杆
overtravel	切出量
overwashing	过洗
oxalic anodic oxidation	草酸阳极化
oxidation	氧化
oxidizing atmosphere	氧化气氛
oxygen (O)	氧
oxygen – pressure ageing	氧压老化
ozone ageing	臭氧老化

P

packaging	包装
packing material	填充/密封材料
paint	涂料
paint sealing	漆封
paint stripper resistance	耐脱漆剂性
pan head	盘头
parallel length	平行长度
parallel pin	圆柱销
parallel thread	圆柱螺纹
parallelism	平行度
parameter	参数
parent metal	母材金属
parent phase	母相
part/product identification number（PIN）	零件识别码
partial annealing	不完全退火
particle size	粒度
parts information model	零件信息模型
passing through type baking oven	通过式烘干室

passivation	钝化
patina	铜绿
peak current	峰值电流
pearlite/pearlyte	珠光体
peeling	剥离/去皮/起皮
peeling strength	剥离强度
peening	喷丸
pencil hardness test	铅笔硬度试验
penetrant flaw detection	渗透探伤
pentagon	五角
penumbra	半影
percent brittle fracture	脆性断裂百分率
percent ductile fracture	塑性断裂百分率
percent shear fracture	剪切断裂百分率
percentage creep elongation	蠕变伸长率
percentage elongation	伸长率
percentage elongation after creep rupture	蠕变断后伸长率
percentage elongation of stress – rupture	持久断后伸长率
percentage extension	延伸率
percentage initial plastic elongation	初始塑性伸长率
percentage permanent elongation	残余伸长率
percentage reduction of area	断面收缩率

percentage total extension at fracture	断裂总延伸率
percentage yield point extension	屈服点延伸率
performance	绩效
performance provision	性能条款
period of validity	有效期
periodic reverse plating	周期转向电镀
permanent deformation/set	永久变形
permeability	渗透性
permeability coefficient	渗透系数
permeation	渗漏/渗透
pH meter	pH 计
pH value	pH 值
phase	相
phase analysis	相位分析
phase transition	相变
phenolic resin	酚醛树脂
phosphating	磷酸盐/磷化处理
phosphorus (P)	磷
photoelastic test	光弹试验
photoelasticity	光弹性
photoelectric absorption	光电吸收
photolithography processing	光刻加工

phrase	相
physical absorption	物理吸附
physical vapor deposition (PVD)	物理气相沉积
pickling	酸洗
piercing	冲孔
pigment	颜料
pilot pin	导正销
pilot point	导向端
pin	销/高锁螺栓
pinhole	针孔
pipe plug	管塞
pipe thread	管螺纹
pit	麻点/坑
pit furnace	井式炉
pitch	螺距
pitch diameter	螺纹中径
pitting corrosion	点蚀
plain washer	平垫圈
plane-strain fracture toughness	平面应变断裂韧度
planing machine/planer	刨床
plasma arc machining (PAM)	等离子弧加工
plasma spraying	等离子喷涂

plastic deformation	塑性变形
plastic elongation time	塑性伸长时间
plastic strain	塑性应变
plastic strain ratio	塑性应变比
plasticate	塑炼
plasticity	塑性
plasticizer	增塑剂
plastics	塑料
plate	板材/厚板（料）
platen press	平板硫化机
plucking	剥胶
pneumatic jigs/fixtures	气动夹具
pockmark	麻点/痘痕
Poisson's ratio	泊松比
polarization	极化
policy	方针
polishing	抛光
Polyethylene (PE)	聚乙烯
polymer	聚合物
polymeric material	聚合材料
polymerization	聚合
Polytetrafluoroethylene (PTFE)	聚四氟乙烯

polyurethane topcoat	聚氨酯面漆
polyvinyl chloride (PVC)	聚氯乙烯
porosity	孔隙率
portable	便携式
position	工位
position tolerance	位置公差
positioning	定位
positive electrode	正极
positive temperature coefficient (PTC)	正温度系数
post emulsification	后乳化
post standard	岗位标准
postplating	镀后处理
potential energy	势能
potential – current density curve	电位-电流密度曲线
powder metallurgy	粉末冶金
power	功(率)
precipitation	析出/沉淀
precipitation hardening stainless steel	沉淀硬化不锈钢
precision machining	精密加工
pre – cleaning	预清洗
precure	预硫化
preferred orientation	最佳取向

preforming	预成形
preheating	预热
preload	预载
preparation	编制
preplating	预镀
press fit sleeve	压配合衬套
press fitting	压装
press riveting	压铆
pressure difference	压力差
pressure feed type spray gun	压送式喷枪
pressure mark	压痕
pressure rising rate	压升率
pressworking	压力加工
prestandard	试行标准
pre – strain	预应变
pre – stress	预应力
pretreatment	预处理
prevailing torque type all metal nut	有效力矩型全金属锁紧螺母
preventive action	预防措施
prevulcanization	预硫化
primary alpha	初生 α 相

primary metalworking	初次成形加工
primer coating	涂底漆
probe	探头/探针
procedure	工艺规程/程序
procedure sheet	工艺过程卡片
process	工艺过程
process accompanying figure	工艺附图
process allowance	工艺余量
process capability index	工序能力系数
process control	过程控制
process control document（PCD）	过程控制文件
process data	工艺数据
process datum	工艺基准
process decision	工艺决策
process design/planning	工艺设计
process dimension	工艺尺寸
process factor	工艺要素
process flow sheet	工艺流程图
process information model	工艺信息模型
process instructions	工艺守则
process optimization	工艺过程优化
process parameter	工艺参数

process program	工艺方案
process route	工艺路线
process sheet	工艺卡片
process specification	工艺规范
process standard	过程标准
process verification	工艺验证
product configuration information	产品技术状态信息
product realization standard	产品实现标准
product standard	产品标准
production batch	生产批量
production cycle	生产周期
production process	生产过程
production program	生产纲领
production time per piece	单件工时
profile	牙型
profile of thread	螺纹牙型
profile plane	侧面
program evaluation and review technique (PERT)	计划评审技术
project	项目
project management	项目管理
project management plan	项目管理计划

projection area	投影区域
proof load	保证载荷
proof load test	保证载荷试验
protective atmosphere	保护气氛
protective material	防护材料
protruding head	凸头
provider/supplier	供方
provincial standards	地方标准
provision	条款
publicly available specification (PAS)	可公开获得的规范
pull type blind rivet	拉拔式抽芯铆钉
pulse	脉冲
pulse anodizing	脉冲阳极化
pulse induction heating equipment	脉冲感应加热装置
pulse load	脉冲载荷
pulse plating	脉冲电镀
punching	冲孔
push broaching	推削
pusher furnace	推送式炉
push - out	推出
pyrometer	高温计
pyrometry	测高温

Q

qualification process	鉴定过程
qualified products list (QPL)	合格产品目录
qualitative	定性的
quality	质量
quality audit	质量审核
quality assurance (QA)	质量保证
quality assurance provision	质量保证条款
quality characteristic	质量特性
quality checking	质量检查
quality control (QC)	质量控制
quality function deployment (QFD)	质量功能展开
quality improvement	质量改进
quality information	质量信息
quality management (QM)	质量管理
quality management system	质量管理系统
quality manual	质量手册
quality objective	质量目标
quality plan	质量计划

quality planning	质量策划
quality policy	质量方针
quality requirement	质量要求
quantitative	定量的
quantitative inspection	计量检验
quenchant	淬火介质
quenched and tempered steel	调质钢
quenching	淬火
quenching tank	淬火槽
quotation	报价

R

raceway	滚道
rack plating	挂镀
radial	径向的
radial clearance	径向游隙
radial ball bearing	向心球轴承
radial bearing	向心轴承
radial deformation	径向变形
radial floating	径向游动
radial permissible static load	径向额定静载荷
radial runout	径向跳动
radial spherical plain bearing	向心关节轴承
radial ultimate static load	径向极限静载荷
radiation	辐射
radius of crest	牙顶圆弧半径
radius of root	牙底圆弧半径
raised cheese head	球面圆柱头
raised countersunk head	半沉头
ram	滑枕

range	范围
rapid prototyping manufacturing（RPM）	快速原型制造
rare earth metal	稀土金属
rare element	稀有元素
ratchet wrench	棘轮扳手
rated output	产量定额
rated pressure	额定压力
raw material	原材料
raw rubber	生胶
reaffirm	重新确认
reamers	铰刀
reaming	铰孔/铰削
recess depth	槽深
recommendation	推荐
record	记录
recrystallization annealing	再结晶退火
recrystallize/recrystallization	再结晶
redox potential	氧化-还原电位
reduced shank	细杆
reducing agent	还原剂
reducing atmosphere	还原气氛
reduction of area	断面收缩率

reference block	标准块
reference line method	基准线法
reference marks	标记/标线
reference plane	参照平面
reference to standards	引用标准
refined steel	精炼钢
refiner	精炼机
refining	精炼
reflection	反射
refraction	折射
regional standard	区域标准
regional standardizing organization	区域标准化组织
regionally harmonized standards	区域协调标准
regression analysis	回归分析
regulation	法规
regulatory requirement	法规要求
reinforcing bead	加强筋
relative humidity	相对湿度
release	放行
release agent	脱模剂
reliability	可靠性
relief hole	排料孔

relieving	铲削
repair	返修
repeatability	重复性
replication	复制
reprint	重新版本
reproducibility	再现性
required elements	必备要素
requirement	要求
residual load	残余载荷
residual magnetic method	剩磁法
residual magnetism	剩磁
residual/internal stress	残余应力
resilience	回弹性
resin	树脂
resistance furnace	电阻炉
resistance heating	电阻加热
resistance welding	电阻焊
resolution	分辨率
resonance curve	共振曲线
resource sharing	资源共享
retaining ring	挡圈
retention capability	保持能力

reticulation	网纹
reverse torsion test of wire	线材反向扭转试验
review	评审/复审
review of technological efficiency	工艺性审查
revision	修订
rework	返工
ribbed cruciform recess	带肋十字槽
right－hand thread	右旋螺纹
rigidity	刚度/刚性
ring	环形件
ring expanding test of tube	管环扩口试验
rinse	清洗/水洗
rinsing equipment	清洗设备
risk	风险
rivet	铆钉
rivet setting load	铆接力
riveting	铆接
riveting facility	铆接设备
rivetless plate nut	无铆托板螺母
Rockwell hardness（HR）	洛氏硬度
rod	条料/杆
rod end spherical plain bearing	杆端关节轴承

rod ends	杆端体
roll coating	辊涂
roll forging	辊锻
roller	滚子
roller bearing	滚子轴承
roller hearth furnace	辊底式炉
rolling	轧制/滚压
rolling bearing	滚动轴承
rolling quenching	旋转淬火
room temperature	室温
room - temperature vulcanization	室温硫化
root	牙底
root mean square (rms)	均方根
root - mean - square (rms) stress	均方根应力
rotary hearth furnace	转底式炉
rotary metalworking	金属回转加工
rough machining	粗加工
rough turning	粗车(削)
roughening	粗化
roughing	粗加工
roughness	粗糙度
round head	圆头

rounded end	倒圆端
rounding/filleting	倒圆角
roundness	圆度
rubber	橡胶
running accuracy	旋转精度
runout	跳动
rust	铁锈
rust preventive oil	防锈油
rust removal	除锈

S

sacrificial anode	牺牲阳极
SAE – Aerospace Information Report (SAE – AIR)	国际自动机工程师协会-航空信息报告
SAE – Aerospace Recommended Practice (SAE – ARP)	国际自动机工程师协会-航空推荐规程
salt bath	盐浴
salt bath deposition	盐浴沉积
salt bath furnace	盐浴炉
salt bridge	盐桥
salt spray test	盐雾试验
sample preparation	试样制备
sampling inspection	抽样检验
sand and dust test	沙尘试验
sand blasting	喷砂
saturation	饱和
sawing machine	锯床
scale	氧化皮
scanning electron microscope (SEM)	扫描电镜

scorch	焦烧
score	划痕/刮痕
scrap	废品/废料/报废
scrap rate	废品率
scraping	刮削
scratch	刮痕/刮伤
scratch hardness	划痕硬度
screen	滤网
screw	螺钉
screw plug	螺塞
screw thread	螺纹
screw thread pair	螺纹副
screw type blind rivet	螺纹型抽芯铆钉
scriber	划针
seal	密封(圈)
sealant	密封胶
sealed bearing	密封圈轴承
sealed box type quenching furnace	箱式淬火炉
sealing	封闭
sealing cap	密封盖
sealing ring	密封圈
seam welding	线焊

seating torque	安装力矩
second annealing	二次退火
secondary metalworking	二次成形加工
segregation	偏析
selective annealing	局部退火
selective laser sintering (SLS)	选择性激光烧结
selenium (Se)	硒
self – aligning bearing	调心轴承
self – aligning washer	自调整垫圈
self – clamping jigs/fixtures	自夹紧夹具
self – locking torque	自锁力矩
self – lubricating bearing	自润滑轴承
self – lubricating bearing bush	自润滑衬套
semi – automatic riveting	半自动铆接
semi – finished product/goods	半成品
sensitivity	灵敏度
sensitization	敏化
sensor	传感器
separable ball bearing	分离式球轴承
service life	使用寿命
service standard	服务标准
servo system	伺服系统

set screw	紧定螺钉
setup	安装
shaft	轴
shaft bracket	轴座
shaft – basis system	基轴制
shaker hearth furnace	振底式炉
shank	钉杆
shape memory effect（SME）alloy	形状记忆合金
shaping	刨削/成形
shatter crack	发裂
shaving	整修/修边
shear	剪切
shear modulus	剪切模量
shear strength	抗剪/剪切强度
shear stress	剪切应力
sheet	薄板(料)/板料
shelf ageing	贮存老化
shelf life	贮存期限/寿命
shield	防尘盖
shielded bearing	防尘盖轴承
Shore hardness	邵氏硬度
short annealing	快速退火

short dog point	短圆柱端
shot	注射量
shot blasting/peening	喷丸
shoulder	轴肩
shrinkage	收缩
shrinkage fitting	热装
sideway	导轨
silicon (Si)	硅
silver (Ag)	银
simulation	仿真
simulative corrosion test	模拟腐蚀试验
single row bearing	单列轴承
single shear test	单剪试验
single - shear joint	单剪接头
single - start thread	单线螺纹
size	尺寸
size tolerance	尺寸公差
skim coating	贴胶
skin annealing	表面退火
sleeve	衬套/钉套
slide block	滑块
sliding	滑动

sling	吊具
slot	开槽
slot depth	开槽深度
slotted capstan screw	开槽带孔球面圆柱头螺钉
slotted cheese head screw	开槽圆柱头螺钉
slotted countersunk head screw	开槽沉头螺钉
slotted headless screw with flat chamferred end	开槽无头倒角端螺钉
slotted pan head screw	开槽盘头螺钉
slotting	插削
slotting machine/slotter	插床
snap ring	止动环
soak heating	均匀加热
soaking	保温
Society of Automotive Engineers (SAE)	国际自动机工程师协会
socket wrench	套筒扳手
solid film lubricant	干膜润滑剂
solid rivet	实心铆钉
solubility	溶解度
soluble developer	可溶显像剂

solution	固溶
solution annealing	固溶退火
solution hardening	固溶硬化
solution heat treatment	固溶热处理
solvent	溶剂
solvent degreasing	有机溶剂除油
soot	炭黑
spalling	剥离
specific gravity	比重
specific gravity cup	比重杯
specific strength/strength – to – weight ratio	比强度
specification	规范
specification standard	规范标准
specimen	试样
spectrometer	光谱仪
spherical outside surface	外球面
spherical plain bearing	关节轴承
spherical plain bearing with self – lubricating liner	自润滑关节轴承
spherical raceway	球面滚道
spherical roller bearing	球面滚子轴承
spheroidizing annealing	球化退火

spindle speed	主轴转速
spinning	旋压
spline	花键
split layer	分层
split pin	开口销
spot corrosion	点蚀
spot repair	局部修补
spot welding	点焊
spotting	锪削
spray booth	喷漆室
spray booth of side exhaust	侧抽风喷漆室
spray coating	喷涂
spray nozzle	喷嘴
spray rinsing	喷洗
spring ring	弹簧圈/卡环
spring washer	弹簧垫圈
square	四方
square head	方头
square neck	方颈
squeeze riveting	压铆
stability	稳定性
stabilized steel	稳定(化)钢

stabilizer	稳定剂
stabilizing annealing	稳定化退火
stabilizing treatment	稳定化处理
stain	污染
stainless steel	不锈钢
stakeholder	相关方
stamping	冲压(件)
stamping stroke	冲压行程
standard	标准
standard compound	标准混炼胶
standard hydrogen electrode	标准氢电极
standard on data to be provided	数据待定标准
standard system	标准体系
standard test block	标准试块
standard time	时间定额
standard tolerance grades	标准公差等级
standard tooling	标准工装
standardization	标准化
standardizing body	标准化机构
standardizing document	标准化文件
standardizing organization	标准化组织
standardizing technical organization	标准化技术组织

standards body	标准机构
standards program	标准工作计划
standards project	标准项目
standoff	紧密距
state of the art	最新技术水平
static analysis	静态分析
static balance test	静平衡试验
static fatigue	静态疲劳
static friction	静摩擦
static load	静载荷
static measurement	静态检测
static strain	静态应变
station	工位
statistical process control（SPC）	统计过程控制
statutory requirement	法定要求
steam sealing	水蒸气封孔处理
steel ball	钢球
steel – steel radial plain spherical bearing	钢对钢向心关节轴承
step	工步
stiffener	硬化剂
stimulation	刺激
stirring equipment	搅拌装置

stock number	物料编码
storage life	贮存期限/寿命
storage stability	贮存稳定性
straight knurl	直纹滚花
straight thread	圆柱螺纹
straightening	矫直
straightening machine	矫直/校直机
straightness	直线度
strain	应变
strain aged steel	应变时效钢
strain age - hardening	应变时效硬化
strain amplitude	应变幅
strain hardening exponent	应变硬化指数
strain rate	应变速率
strain - hardening	应变硬化
strategy	战略
strength of a riveted joint	铆接接头强度
strengthening	强化
stress	应力
stress amplitude	应力幅
stress area	应力面积
stress corrosion	应力腐蚀

stress durability	应力持久
stress intensity factor	应力强度因子
stress rate	应力速率
stress relaxation	应力松弛
stress relaxation curve	应力松弛曲线
stress relaxation rate	应力松弛速率
stress relief annealing	去应力退火
stress relieving	去应力退火
stress rupture	应力断裂
stress yield	应力屈服
stress – rupture limit	持久强度极限
stress – rupture notch sensitivity factor	持久缺口敏感系数
stress – strain curve	应力-应变曲线
strike/underplating	打底
strip	带料
stripping	脱膜/退镀
stroke	行程
structural panel fastener lap joint shear test	结构用托板紧固件搭接接头剪切试验
structural steel	结构钢
structure	结构/组织
stud	螺柱

stud with undercut	带退刀槽螺柱
styrene – butadiene rubber	丁苯橡胶
subboundary	亚晶界
subcommittee (SC)	分技术委员会
subcritical austenite	亚临界奥氏体
sub – grain	亚晶粒/次级晶粒
subject of standardization	标准化对象
substrate	基体材料/底材
substructure	亚组织
sulfur (S)	硫
sulfuric acid	硫酸
sulphoacid anodic oxidation	硫酸阳极化
superfinishing	超精加工
surface active agent/surfactant	表面活性剂
surface contamination	表面污染
surface discontinuity	表面不连续
surface grinding	平面磨
surface hardening	表面淬火
surface heat treatment	表面热处理
surface roughness	表面粗糙度
surface texture	表面结构
surface treatment	表面处理

suspension	磁悬液
swaged collar type rivet	环槽铆钉
swaging	模锻
swaging groove	安装槽
sweep range	扫描范围
sweep speed	扫描速度
sweep/scan	扫描
swellability	溶胀度
swelling	溶胀
swivel joint	旋转接头
swivelling/tilting torque	偏转力矩
symbol standard	符号标准
symmetry	对称性
synthetic rubber	合成橡胶
system structure diagram	体系结构图

T

tab washer with long tab	单耳止动垫圈
table	工作台
tangential test	切向取样试验
taper pin	圆锥销
taper thread	圆锥螺纹
taper washer	斜垫圈
tapping	攻丝
tapping screw	自攻螺钉
tapping thread	自攻螺纹
target	靶
tarnish	金属变色
tear strength	撕裂强度
technical committee（TC）	技术委员会
technical deviation	技术性差异
technical expert	技术专家
technical normative elements	规范性技术要素
technical regulation	技术法规
technical report（TR）	技术报告

technical requirement	技术要求
technical specification (TS)	技术规范
technical standard	技术标准
technical standard system	技术标准体系
technological documentation	工艺文件
technological management	工艺管理
technological materials	工艺材料
technological preparation of production	工艺准备
technology	工艺
technology of metals	金属工艺学
temper ageing	回火时效
temper brittleness	回火脆性
temperature calibration	温度标定
temperature of phase change	相变温度
temperature stress	温差应力
temperature uniformity	温度均匀性
tempering	回火
tensile fatigue	拉伸疲劳
tensile modulus	拉伸模量
tensile set	拉伸永久变形
tensile strength	抗拉/拉伸强度
tensile stress	拉伸应力

tensile stress area	螺纹抗拉强度面积
tensile stress at a given elongation	定伸应力
tensile stress at yield	屈服拉伸应力
tensile test	拉伸试验
tensile test under wedge loading	楔负载拉力试验
tension bar	拉杆
tension test	拉力试验
terminology standard	术语标准
test	试验
test block	试块
test of resistance to high temperature oxidation	抗高温氧化试验
test panel	试验板
test specimen/piece	试样/试件
testing standard	试验标准
texture	织构/纹理
T – head	T 形头
theoretical curing time	理论正硫化时间
thermal conductivity	热导率
thermal degradation	热降解
thermal diffusion	热扩散
thermal diffusivity	热扩散率

thermal image/thermogram	热像图
thermal imaging system	热成像系统
thermal resolution	热分辨率
thermal shock test	热冲击试验
thermal spraying	热喷涂
thermal strain	热应变
thermal stress	热应力
thermo – chemical treatment	化学热处理
thermocouple	热电偶
thermo – intensity of hearth	炉底热强度
thermo – mechanical treatment（TMT）	形变热处理
thermoplastic rubber	热塑性橡胶
thermostability	热稳定性
thickness variation	壁厚差
thickness/plug gage	塞尺
thin wall self – locking insert	薄壁自锁螺套
thin – walled bearing	薄壁轴承
third angle projection	第三角投影
thread	螺纹
thread angle	牙型角
thread indicating gauge	螺纹指示规
thread machining	螺纹加工

thread plug gauge	螺纹塞规
thread ring gauge	螺纹环规
thread rolling	滚丝
thread runout	螺纹收尾
thread shear area	螺纹抗剪强度面积
thread snap gauge	螺纹卡规
thread start	引导螺纹
thread tools	螺纹刀具
threads per inch (t. p. i)	牙数
thread - setting gauge	螺纹定位校对规
three dimensional printing	3D(三维)打印
three point contact ball bearing	三点接触球轴承
threshold	临界点/值
threshold strain	临界应变
throwing power	分散能力
thust ball bearing	推力球轴承
thust bearing	推力轴承
tight fit	紧配合
tightening torque	拧紧力矩
time synchronous display	时间同步显示
tin (Sn)	锡
tinting strength	着色强度

titanium（Ti）	钛
titanium alloy	钛合金
titration solution	滴定溶液
tolerance	公差
tolerance class	公差代号
tolerance grade	公差等级
tolerance position	公差带位置
tolerance zone	公差带
tool box/kit	工具箱
tool post/rest	刀架
tool retracting	退刀
tooling	工艺装备/工装
tools holder	刀夹
top view	俯视图
torque	力矩/扭矩
torque tension test	扭矩-拉力试验
torque wrench	扭力扳手
torque – out	扭出
torsion	扭转
torsion test of wire	线材扭转试验
torsional strength	抗扭强度
total acid	总酸度

total allowance for machining	加工总余量/毛坯余量
total quality control	全面质量管理
tough fracture	韧性断口
toughening	韧化(处理)
toughness	韧度/韧性
toxicity	毒性
traceability	可追溯性
transformation annealing	相变退火
transformation embrittlement	相变脆性
transformation stress	相变应力
transgranular cracking	穿晶断裂
transgranular stress corrosion cracking	应力腐蚀穿晶断裂
transient surface	过渡表面
transition fit	过渡配合
transition temperature	转变温度
transport number	迁移数
transversal strain	横向应变
transverse resolution	横向分辨率
transverse/longitudinal flux heating	横向/纵向磁通加热
travel	行程
trial assembly	试装

triangle	三角
trimming	整修/修边
troostite	屈氏体
true strain	真应变
true stress	真应力
truncated cone point	截锥端
tubing	管材
tubular rivet	空心铆钉
tumble polishing	滚筒抛光
tungsten（W）	钨
turbidity	浊度
turbine engine	涡轮发动机
turnbuckle	螺旋扣
turning	车削
turning machine	车床
twin	孪晶
twin boundary	孪晶边界
two lug plate nut	双耳托板螺母
type test	型式试验

U

ultimate tensile strength（UTS）	极限抗拉强度
ultra high frequency induction heating equipment	超高频感应加热装置
ultraprecision machining	超精密加工
ultrasonic cleaning	超声波清洗
ultrasonic flaw detection	超声探伤
ultrasonic machining（USM）	超声加工
ultrasonic microscope	超声显微镜
ultrasonic welding	超声波焊
ultrasound EDM	超声电火花加工
ultrasound grinding	超声研磨
ultraviolet ray	紫外线
undated reference（to standards）	（对标准的）不注日期引用
underhead fillet	头下圆角
underhead radius	头下半径
uniaxial	单轴的
uniaxial stress	单向应力

unified screw thread	统一螺纹
unified standards	一致标准
uniform corrosion	均匀腐蚀
uniformity	均匀性/一致性
uniphase	单相(的)
universal tensile testing machine	万能拉伸试验机
universal tooling	通用工装
unloading	卸料
U – notch	U 型缺口
unseating torque	拆卸力矩
untightening torque	拧松力矩
upper limit deviation（ES）	上偏差
upper yield strength	上屈服强度
upsetting	镦锻/镦粗/顶镦
useful thread	有效螺纹

V

vacuum	真空
vacuum annealing	真空退火
vacuum arc remelting（VAR）	真空电弧重熔
vacuum cassette	真空暗盒
vacuum deposition	真空沉积
vacuum furnace	真空炉
vacuum hardening/quenching	真空淬火
vacuum heat treatment	真空热处理
vacuum resistance furnace	真空电阻炉
vacuum tempering	真空回火
validation	确认
vanadium（V）	钒
vanish thread	螺纹收尾
vapor degreasing	蒸汽除油
varnish	清漆
verification	检验/验证
vernier caliper	游标卡尺
vertical vulcanizer	立式硫化机

vibration	振动
vibration test	振动试验
vibrocutting	振动切削
vice	虎钳
vice versa principle	反之亦然原则
Vickers hardness（HV）	维氏硬度
view	视图
virgin rubber	原胶
virtual assembly	虚拟装配
virtual manufacturing	虚拟制造
viscosity	粘度
visible light	可见光
vision	愿景
visual examination	外观检查
V‒notch	V 型缺口
void	空洞
volatile fatty acid number	挥发性脂肪酸值
volume current density	体积电流密度
volume density	体积密度
vulcanization	硫化
vulcanized rubber	硫化橡胶
vulcanizing agent	硫化剂

W

waisted shank	腰状杆
waisted stud	腰状杆螺柱
walking beam furnace	步进式炉
warm forging	温锻
warping	翘曲
washer	垫圈
washout thread	螺纹收尾
Waspaloy	耐高热镍基合金
water break	水膜残迹/不连续水膜
water quenching	水淬
water toughening	水韧处理
water - washable penetrant	可水洗型渗透剂
wave	波
wave washer	波形垫圈
waviness	波纹度
wear	磨耗/磨损
weathering	气候老化/风化
wedge tensile	楔垫拉伸

wedge testing	楔形块试验
weld nut	焊接螺母
welding	焊接（件）
wet assembly	湿装配
wet chemistry method	湿化学法
wet spots	湿斑
wetting agent	润湿剂
wheel nut	车轮螺母
white haze	白雾
white spots	白斑
Whitworth thread	惠氏螺纹
wing	蝶/翼形
wing screw	翼形螺钉
wire	线材/丝材
wire drawing	拉丝
wiredraw blind rivet	拉丝型抽芯铆钉
wood screw	木螺钉
work surface	待加工表面
working gauge	工作规
working group（WG）	工作组
working pressure	工作压力
working temperature	工作温度

working/operating stroke	工作行程
work – in – process	在制品
workpiece	工件
World Intellectual Property Organization（WIPO）	世界知识产权组织
wrapping test	缠绕试验
wrapping test of wire	线材缠绕试验
wrench	扳手
wrench torque	扳拧力矩
wrenching allowance	扳紧余量
wrenching configuration	扳拧结构
wrenching feature	扳拧特征
wrinkling	起皱

X

| X – ray detection apparatus | X 射线探伤装置 |
| X – ray fluorescence（XRF） | X 射线荧光 |

Y

yield strength	屈服强度
yield stress	屈服应力
yoke	磁轭
yoked coil	轭式线圈
Young's modulus	杨氏模量
yttrium（Y）	钇

Z

zero defect	零缺陷
zero line	零线
zinc (Zn)	锌
zinc dipping baths	热浸镀锌浴
zinc rich primer	富锌底漆
zonal structure	带状组织
zonal texture	带状织构
zone location	区域定位
zone of columnar crystal	柱状晶区/晶带
zone of equiaxed crystal	等轴晶区/晶带
zone of interaction	作用区
zone of plastic deformation	塑性变形区
ГОСТ	俄罗斯国家标准

中文→英文

产品与结构
Products & Features

凹端	cup point
八角	octagon
八角头	octagonal head
扳紧余量	wrenching allowance
扳拧结构	wrenching configuration
扳拧特征	wrenching feature
扳拧形状	driving feature
扳手	wrench
半沉头	oval head/raised countersunk head
包络螺纹	enveloping thread
边距	edge distance
扁圆头	cup head/mushroom head
扁圆头带榫螺栓	mushroom head nib bolt
标记	marking
标识/标志	identification
标准杆	normal shank
标准公差等级	standard tolerance grades
波浪线	break line

波形垫圈	wave washer
薄壁自锁螺套	thin wall self – locking insert
不完整螺纹	incomplete thread
参照平面	reference plane
槽深	recess depth
槽销	grooved pin
槽型螺母	gang channel nut
侧面	profile plane
产品/项（目）	item
长圆柱端	long dog point
车轮螺母	wheel nut
沉头	countersunk head
沉头铆钉	flush rivet
衬垫	gasket
衬套	bush/sleeve
成组游动托板自锁螺母	gang channel floating anchor self – locking nut
承力快卸锁	bearing quick release latch
承载牙侧	load flank
尺寸	size
尺寸公差	dimensional/size tolerance
抽芯铆钉/盲铆钉	blind rivet

粗牙螺纹	coarse pitch thread
寸制螺纹	inch thread
大底脚螺纹型抽芯铆钉	large footprint thread – type blind rivet
带截断心轴的抽芯铆钉	blind rivet with break mandrel
带肋十字槽	ribbed cruciform recess
带退刀槽螺柱	stud with undercut
单耳止动垫圈	tab washer with long tab
单剪接头	single – shear joint
单面间隙	blind side clearance
单线螺纹	single – start thread
挡圈	retaining ring
导程	lead
导程角/升角	lead angle
导向端	pilot point
导正销	pilot pin
倒角端	chamfered end
倒圆端	rounded end
等长双头螺柱	double end stud
低圆柱头	cheese head
底座	base

地脚螺栓	foundation bolt
第三角投影	third angle projection
垫圈	washer
蝶形/翼形	wing
钉杆	shank
钉套	sleeve
钉体	body
钉芯	mandrel
定位零件	locating elements
定位销	locating pin
短圆柱端	short dog point
对边宽度	across flats
对角宽度	across corners
多线螺纹	multi – start thread
法兰	flange
方颈	square neck
方头	square head
封闭型抽芯铆钉	closed – end blind rivet
俯视图	top view
副产品	by – products
盖形螺母	acorn nut
杆部形状	forms of shank

高锁螺母	high-locking/hi-lock nut/collar
高锁螺栓	high-locking/hi-lock bolt/pin
工作规	working gauge
公差	tolerance
公差代号	tolerance class
公差带	tolerance zone
公差带位置	tolerance position
公差等级	tolerance grade
公称尺寸	nominal size
公称直径	nominal diameter
鼓包型抽芯铆钉	bulbed blind rivet
关键零件	critical part
（管）接头	coupler/coupling
管螺纹	pipe thread
管塞	pipe plug
滚花	knurl
滚花高头螺钉	knurled thumb screw
过渡配合	transition fit
过盈/干涉配合	interference fit
焊接螺母	weld nut
合格	conformity
合格品	conforming

花键	spline
滑块	slide block
环槽铆钉	swaged collar type rivet
环形件	ring
惠氏螺纹	Whitworth thread
活节螺栓	eye bolt
击入式钉芯	drive pin
机身	frame
基本尺寸	basic size
基本偏差	fundamental deviation
基本三角形	fundamental triangle
基本牙型	basic profile
基准/数据	datum
基准长度/标距	gauge length
基准平面	gauge plane
基准直径	gauge diameter
极限尺寸	limits of size
极限偏差	limit deviation
棘轮扳手	ratchet wrench
加大杆	oversize shank
加强杆	fit shank/increased shank
夹层长度	grip length

夹层范围	grip range
间隙	clearance
间隙配合	clearance fit
减轻孔直径	lightening hole diameter
接头	joint
结构	structure
截锥端	truncated cone point
紧定螺钉	set screw
紧公差	close tolerance
紧固件	fastener
紧密距	standoff
紧配合	tight fit
径向游动	radial floating
均方根	root mean square（rms）
卡箍	clamp
卡箍螺栓	clip bolt
开槽	slot
开槽沉头螺钉	slotted countersunk head screw
开槽带孔球面圆柱头螺钉	slotted capstan screw
开槽盘头螺钉	slotted pan head screw
开槽深度	slot depth

开槽无头倒角端螺钉	slotted headless screw with flat chamferred end
开槽圆柱头螺钉	slotted cheese head screw
开叉地脚螺栓	masonry bolt
开口销	cotter pin/split pin
开口型沉头抽芯铆钉	open‑end blind rivet with countersunk head
空心铆钉	hollow‑core rivet/tubular rivet
口盖	hatch cover
口盖锁	cover latch
拉拔式抽芯铆钉	pull type blind rivet
拉丝型抽芯铆钉	wiredraw blind rivet
联杆销	link rod pin
零线	zero line
六角	hexagon
六角薄螺母	hexagon thin nut
六角垫圈面头	hexagon head with washer face
六角法兰面螺栓	hexagon head bolt with flange
六角法兰面头	hexagon head with flange
六角花形	hexalobular
六角花形头	hexalobular head
六角皇冠螺母	hexagon castle nut
六角头	hexagon head

六角头螺塞	hexagon head screw plug
六角凸缘螺栓	hexagon head bolt with collar
六角凸缘头	hexagon head with collar
螺钉	screw
螺距	pitch
螺距偏差	deviation in pitch
螺母	nut
螺塞	screw plug
螺栓	bolt
螺套/嵌件	insert
螺纹	(screw) thread
螺纹大径	major diameter
螺纹定位校对规	thread – setting gauge
螺纹副	screw thread pair
螺纹环规	thread ring gauge
螺纹卡规	thread snap gauge
螺纹抗剪强度面积	thread shear area
螺纹抗拉强度面积	tensile stress area
螺纹塞规	thread plug gauge
螺纹收尾	thread runout/vanish thread/washout thread
螺纹小径	minor diameter
螺纹型抽芯铆钉	screw type blind rivet

螺纹牙型	form/profile of thread
螺纹指示规	thread indicating gauge
螺纹中径	pitch diameter
螺纹轴线	axis of thread
螺旋扣	turnbuckle
螺旋线	helix
螺旋线导程	lead of helix
螺柱	stud
MJ 螺纹	MJ profile thread
盲螺栓	blind bolt
锚栓	anchor bolt
铆钉	rivet
米制螺纹	metric thread
密封盖	sealing cap
密封圈	sealing ring
面板螺钉	cover screw
末端	end
木螺钉	wood screw
内径	inside diameter (ID)
内六角	hexagon socket
内六角圆柱头螺钉	hexagon socket head cap screw
内螺纹	internal thread

尼龙圈	nylon washer
扭力扳手	torque wrench
盘头	pan head
配合	fit
喷嘴	nozzle
膨胀铆钉	expanded rivet
偏差	deviation
偏心十字槽	offset cruciform recess
平垫圈	plain washer
平端	flat point
平面度	flatness
平齐度	flushness
平行度	parallelism
平圆头	button head
球面圆柱头	raised cheese head
全螺纹	full thread
软管	hose
软管接头	fitting
三角	triangle
上偏差	upper limit deviation（ES）
设计牙型	design profile
十二角	bihexagon

十二角头	bihexagonal head
十字槽	cross/cruciform recess
十字槽盘头螺钉	cross recessed pan head screw
十字孔	cross hole
实际尺寸	actual size
实体牙型	material profile
实心铆钉	solid rivet
视图	view
双耳托板螺母	double – lug anchor nut/two lug plate nut
双剪接头	double – shear joint
双金属铆钉	bimetallic rivet
双线螺栓	bilinear bolt
水平面	horizontal plane
四方	square
锁环	locking ring
锁紧垫圈	lock washer
锁紧零件	locking element
锁紧特性	locking feature
T 形头	hammer head/T – head
弹簧垫圈	spring washer
套筒扳手	socket wrench
填芯铆钉	filled – core rivet

条形码	bar code
跳动	runout
通规	GO gauge
同轴度	concentricity
统一螺纹	unified screw thread
头部保险孔直径	head locking hole diameter
头部形状	head shapes
头下半径	underhead radius
头下圆角	underhead fillet
投影区域	projection area
凸头	protruding head
凸缘	collar
推杆	ejector pin
外径	outside diameter（OD）
外螺纹	external thread
外协件	cooperation part
完整螺纹	complete thread
网纹滚花	diamond knurl
位置公差	position tolerance
涡轮发动机	turbine engine
无铆托板螺母	rivetless plate nut
五角	pentagon

误差	error
细杆	reduced shank
细牙螺纹	fine pitch thread
下偏差	lower limit deviation (EI)
销	pin
销轴	clevis pin
校对规	check gauge
斜垫圈	taper washer
行程	kinematic path/travel
旋转接头	swivel joint
压配合衬套	press fit sleeve
牙侧	flank
牙底	root
牙底圆弧半径	radius of root
牙顶	crest
牙顶圆弧半径	radius of crest
牙数	threads per inch (t. p. i)
牙型	form/profile
牙型角	thread angle
仰视图	bottom view
腰状杆	waisted shank
腰状杆螺柱	waisted stud

翼形螺钉	wing screw
引导螺纹	lead thread/thread start
引导牙侧	leading flank
游动托板螺母	floating anchor/plate nut
有效力矩型全金属锁紧螺母	prevailing torque type all metal nut
有效螺纹	effective/useful thread
右旋螺纹	right‑hand thread
圆度	roundness
圆头	round head
圆柱螺纹	cylindrical/parallel/straight thread
圆柱头	cylindrical head
圆柱销	parallel pin
圆锥度/圆柱度	cylindricity
圆锥螺纹	taper thread
圆锥销	taper pin
辗制末端	as‑rolled end
正面	frontal plane
支承面	bearing surface/face
直纹滚花	straight knurl
直线度	straightness
止规	NO GO gauge

中径轴线	axis of pitch diameter
轴	shaft
轴肩	shoulder
转接半径	blend radius
装配余量	fitting allowance
锥端	cone point
自调整垫圈	self – aligning washer
自攻螺钉	tapping screw
自攻螺纹	tapping thread
纵坐标	ordinate
钻出铆钉	drilled – out rivet
最大极限尺寸	maximum limits of size
最大实体极限尺寸	maximum material limit（MML）
最大实体牙型	maximum material profile
最佳螺纹量针或量球	best – size thread wire or ball
最小极限尺寸	minimum limits of size
最小实体极限尺寸	least material limit（LML）
最小实体牙型	minimum material profile
左旋螺纹	left – hand thread

标准化
Standardization

保障标准	assurance standard
必备要素	required elements
必达要求	exclusive requirement
编辑性修改	editorial change
编制	preparation
标准	standard
标准草案	draft standard
标准的间接应用	indirect application of an standard
标准的直接应用	direct application of an standard
标准工作计划	standards program
标准化	standardization
标准化层次	level of standardization
标准化对象	subject of standardization
标准化机构	standardizing body
标准化技术组织	standardizing technical organization
标准化领域	field of standardization
标准化文件	standardizing document
标准化组织	standardizing organization

标准机构	standards body
标准体系	standard system
标准项目	standards project
采用	adoption
产品标准	product standard
产品几何技术规范	geometrical products specifications（GPS）
产品实现标准	product realization standard
德国标准化学会	DIN（Deutsches Institut für Normung）
等同	identical（IDT）
等同标准	identical standards
地方标准	provincial standards
（对标准的）不注日期引用	undated reference（to standards）
（对标准的）注日期引用	dated reference（to standards）
俄罗斯国家标准	ГОСТ
法规	regulation
反之亦然原则	vice versa principle
非等效	nonequivalent（NEQ）
分技术委员会	subcommittee（SC）
分类标准	classification standard

服务标准	service standard
符号标准	symbol standard
复审	review
岗位标准	post standard
工程更改单	engineering change order
工作标准	duty standard
工作标准体系	duty standard system
工作组	working group（WG）
管理标准	administrative standard
管理标准体系	administrative standard system
规程	code of practice
规程标准	code of practice standard
规范	specification
规范标准	specification standard
规范性技术要素	technical normative elements
规范性文件	normative document
规范性文件的采用	adoption of a normative document
规范性文件的应用	application of a normative document
规范性要素	normative elements
规范性一般要素	general normative elements
国际标准	international standard
国际标准化	international standardization

国际标准化组织	International Organization for Standardization（ISO）
国际标准组织	international standards organization
国际电工委员会	International Electrotechnical Commission（IEC）
国际电信联盟	International Telecommunication Union（ITU）
国际焊接学会	International Institute of Welding（IIW）
国际计量局	Bureau International des Poids et Mesures（BIPM）
国际民航组织	International Civil Aviation Organization（ICAO）
国际协调标准	internationally harmonized standards
国际自动机工程师协会	Society of Automotive Engineers（SAE）
国际自动机工程师协会-航空推荐规程	SAE – Aerospace Recommended Practice（SAE – ARP）
国际自动机工程师协会-航空信息报告	SAE – Aerospace Information Report（SAE – AIR）
国家标准	national standard
国家标准化	national standardization

国家标准机构	national standards body
过程标准	process standard
互换性	interchangeability
基础标准	basic standard
技术报告	technical report（TR）
技术标准	technical standard
技术标准体系	technical standard system
技术法规	technical regulation
技术规范	technical specification（TS）
技术委员会	technical committee（TC）
技术性差异	technical deviation
兼容性	compatibility
接口标准	interface standard
结构	structure
勘误	correction
可比标准	comparable standards
可公开获得的规范	publicly available specification（PAS）
可选要求	optional requirement
可选要素	optional elements
（美）电器和电子工程师学会	Institute of Electrical and Electronic Engineers（IEEE）
（美）国家航空标准	National Aerospace Standard（NAS）

(美)国家航空航天和国防合同方授信项目	National Aerospace and Defense Contractors Accreditation Program (NADCAP)
(美)航空标准	Aerospace Standard (AS)
(美)航空材料规范	Aerospace Material Specification (AMS)
(美)航空工业协会	Aerospace Industries Association (AIA)
美国材料与试验协会	American Society for Testing and Materials (ASTM)
美国国家标准学会	American National Standard Institute (ANSI)
美国机械工程师协会	American Society of Mechanical Engineers (ASME)
欧洲标准	EN (Europäische Norm)
批准	approval
企业标准	company standard
企业标准体系	enterprise standard system
起草	drafting
区域标准	regional standard
区域标准化组织	regional standardizing organization
区域协调标准	regionally harmonized standards
认定机构	cognizant activity
认可	accreditation
日本工业标准	Japanese Industrial Standard (JIS)

日本工业标准调查会	Japanese Industrial Standards Committee (JISC)
世界知识产权组织	World Intellectual Property Organization （WIPO）
试行标准	prestandard
试验标准	testing standard
适航	airworthiness
适用性	fitness for purpose
术语标准	terminology standard
数据待定标准	standard on data to be provided
体系结构图	system structure diagram
条款	provision
团体	association
团体标准	association standard
推荐	recommendation
物料编码	stock number
协调标准	harmonized/equivalent standards
协商一致	consensus
新版本	new edition
行业标准	industry standard
性能条款	performance provision
修订	revision

修改	modified（MOD）
修正	amendment
要求	requirement
一致标准	unified standards
一致性程度	degrees of correspondence
引用标准	reference to standards
英国标准学会	British Standards Institution（BSI）
有效期	period of validity
职业安全与卫生条例	occupational safety and health act（OSHA）
指南	guideline
指南标准	guide standard
指示	instruction
制定	development
重新版本	reprint
重新确认	reaffirm
资料性要素	informative elements
最新技术水平	state of the art

机械加工
Machining

安装	installation/setup
摆动	oscillating
半成品	semi – finished product/goods
半自动铆接	semi – automatic riveting
包装	packaging
报价	quotation
变速箱	gearbox
标准工装	standard tooling
剥胶	plucking
补偿	compensating
材料消耗工艺定额	material consumption quota in process
残屑	chip
测量基准	measuring datum
插床	slotting machine/slotter
插孔	hole slotting
插削	slotting
拆卸	disassembly
产量定额	rated output

铲削	relieving
超精加工	superfinishing
超精密加工	ultraprecision machining
超声波焊	ultrasonic welding
超声电火花加工	ultrasound EDM
超声加工	ultrasonic machining（USM）
超声研磨	ultrasound grinding
车床	turning machine
车床夹具	lathe fixture
车孔	hole turning
车削	turning
沉孔	countersink
成品	final product
成形	forming
成形性	formability
成组技术	group technology（GT）
尺寸链	dimensional chain
齿轮刀具	gear cutters
冲裁/落料	blanking
冲裁力	blanking force
冲孔	piercing/punching
冲压(件)	stamping

冲压行程	stamping stroke
初次成形加工	primary metalworking
传感器	sensor
吹塑	blow molding
粗车(削)	rough turning
粗加工	roughing / rough machining
搓丝	flat die thread rolling
搭接(接头)	lap joint
待加工表面	work surface
单件工时	production time per piece
刀杆	arbor
刀夹	tools holder
刀架	tool post/rest
刀尖	corner
刀具	cutting tool
导板	guide plate
导轨	sideway
倒钝锐边	breaking sharp corners
倒角	chamfer
倒圆角	filleting/rounding
等离子弧加工	plasma arc machining (PAM)
等温锻	isothermal forging

点动	inching/jogging
点焊	spot welding
电弧焊	arc welding
电火花加工	electro – discharge machining（EDM）
电解加工	electro – chemical machining（ECM）/ electrolytic machining
电解磨削	electrolytic grinding
电解抛光	electrolytic polishing
电子束加工	electron beam machining（EBM）
电阻焊	resistance welding
调整/调节	adjusting
定位	positioning
定位基准	fixed datum
动态硫化	dynamic vulcanization
端面密封	face seal
锻件/锻造	forging
锻压	forging and stamping
锻造流线	forging flow line
对刀件	elements for aligning tool
镦锻/顶镦	heading/upsetting
多工位自动成型机	multi – station automatic former
二次成形加工	secondary metalworking

仿真	simulation
废品	scrap
废品率	scrap rate
分层实体制造	laminated object manufacturing
分度运动	dividing movement
粉末冶金	powder metallurgy
辅助工步	auxiliary step
复合加工	combined machining (CM)/compound machining
干式切削	dry cutting
高速高能成形	high – energy – rate forming (HERF)
高速切削	high – speed cutting
高压水切割	high pressure water cutting
工步	step
工件	workpiece
工具箱	tool box/kit
工时定额	man – hour quota
工位	position/station
工序	operation
工序尺寸	operation dimension
工序基准	operation datum
工序卡片	operation sheet

工序能力系数	process capability index
工序图	operation diagram
工序余量	operation allowance
工艺	technology
工艺参数	process parameter
工艺尺寸	process dimension
工艺方案	process program
工艺附图	process accompanying figure
工艺管理	technological management
工艺规程	procedure
工艺规范	process specification
工艺过程	process
工艺过程卡片	procedure sheet
工艺过程优化	process optimization
工艺基准	process datum
工艺纪律	manufacturing discipline
工艺决策	process decision
工艺卡片	process sheet
工艺流程图	process flow sheet
工艺路线	process route
工艺设备	manufacturing equipment
工艺设计	process design/planning

工艺试验	engineer test
工艺守则	process instructions
工艺数据	process data
工艺凸台	false boss
工艺文件	technological documentation
工艺文件更改通知单	change order for technological documentation
工艺系统	machining complex
工艺信息模型	process information model
工艺性分析	analysis of technological efficiency
工艺性审查	review of technological efficiency
工艺验证	process verification
工艺要素	process factor
工艺余量	process allowance
工艺装备/工装	tooling
工艺准备	technological preparation of production
工作台	table
工作行程	working/operating stroke
功(率)	power
攻丝	tapping
刮削	scraping
光弹性	photoelasticity
光刻加工	photolithography processing

光整加工	finishing cut
辊锻	roll forging
滚花	knurl
滚丝	thread rolling
滚筒抛光	barrel/tumble polishing
滚压孔	hole rolling
过程控制文件	process control document（PCD）
过渡表面	transient surface
过硫化	overcure
焊接(件)	welding
珩孔	hole honing
珩磨	honing
横梁	cross rail
虎钳	vice
滑动	sliding
滑枕	ram
划针	scriber
混炼	compounding
锪孔	countersinking/counterboring
锪削	counterboring/spotting
机械调节	mechanical conditioning
机械化	mechanization

机械加工	machining
机械加工余量	machining allowance
机械密封	mechanical seal
基材	substrate
激光加工	laser beam machining (LBM)
集成制造系统	integrated manufacturing system
挤出	extrusion
挤压	extruding
计划评审技术	program evaluation and review technique (PERT)
计算机辅助测试	computer – aided test (CAT)
计算机辅助工程	computer – aided engineering (CAE)
计算机辅助工艺设计	computer – aided process planning (CAPP)
计算机辅助设计	computer – aided design (CAD)
计算机辅助制造	computer – aided manufacturing (CAM)
计算机集成制造系统	computer integrated manufacturing system (CIMS)
计算机数控	computer numerical control (CNC)
技术要求	technical requirement
加工精度	machining accuracy
加工误差	machining error
加工中心	machining center (MC)

加工总余量/毛坯余量	total allowance for machining
加强筋	reinforcing bead
加速硫化	accelerated cure
夹具	fixture/jigs
夹头	collet chuck
矫直	straightening
铰刀	reamers
铰孔/铰削	reaming
接杆	extension bar
接口/界面	interface
金属回转加工	rotary metalworking
金属流动	metal flow
金属塑性加工	metal plastic working
金属注射成型	metal injection moulding (MIM)
进给量	feed
进给速度	feed speed
浸渍	dipping
精车(削)	finish turning
精加工	finishing
精炼	refining
精炼机	refiner
精密加工	precision machining

局域网	local area network（LAN）
锯床	sawing machine
均匀变形	homogeneous deformation
卡盘	chuck
卡爪	jaw
可锻性	forgeability
可加工性	machinability
刻线	dividing
空锻	hollow forging
空行程	idle stroke
快速原型制造	rapid prototyping manufacturing（RPM）
扩孔	expanding/counterboring
扩口	flaring
拉拔	drawing
拉床	broaching machine/broacher
拉杆	tension bar
拉孔	hole broaching
拉丝	wire drawing
拉削	broaching/pull broaching
冷拔/冷拉	cold drawing
冷变形强化	cold deformation strengthening
冷锻	cold forging

冷镦	cold heading/upsetting
冷隔	cold shut
冷挤压	cold extrusion
冷加工/冷作	cold work
冷矫直	cold straightening
冷精整	cold finish
冷硫化	cold cure
冷轧	cold rolling
冷装	expansion fitting
离子束加工	ion beam machining (IBM)
理论正硫化时间	theoretical curing time
立式硫化机	vertical vulcanizer
连续混炼机	continuous mixer
连续硫化	continuous vulcanization
连续硫化机	continuous vulcanizer
临界变形程度	critical deformation
临界面	critical plane
零件识别码	part/product identification number (PIN)
零件信息模型	parts information model
流程图	flow chart/diagram
硫化	vulcanization
硫化/固化	cure

硫化罐	autoclave
螺纹刀具	thread tools
螺纹加工	thread machining
盲孔	blind hole
毛刺	burr
毛坯	blank
毛坯图	blank drawing
铆接	riveting
铆接设备	riveting facility
密炼机	internal mixer
敏捷制造	agile manufacturing
模锻	swaging
模锻(件)	die forging
模具	die/mould
模腔/型腔	cavity
模塑/成型	moulding
模压	compression moulding
模压收缩	moulding shrinkage
摩擦焊	friction welding
磨床	grinding machine/grinder
磨孔	hole grinding
磨削	grinding

磨削刀具	abrasive tools
母材金属	base/parent metal
内圆磨	internal cylindrical grinding
纳米加工	nano - processing
排料孔	relief hole
排屑装置	chip conveyor
抛光	polishing
刨床	planing machine/planer
刨削	gouging
刨削/成形	shaping
喷砂	sand blasting
喷丸	peening/shot blasting
批量试制	lot pilot
偏心	eccentricity
平板硫化机	platen press
平面磨	surface grinding
普通机床	general accuracy machine tools
漆封	paint sealing
气动夹具	pneumatic jigs/fixtures
钳工工具	bench - work tool
钳加工	bench work
强化	strengthening

切槽	grooving
切出量	overtravel
切入量	approach
切削功率	cutting power
切削加工	cutting
切削加工工装	cutting tooling
切削力	cutting force
切削温度	cutting temperature
切削液	cutting fluid
清洗	cleaning
球磨机	ball mill
去毛刺	deburring
让刀/抬刀	cutter relieving
热锻	hot forging
热镦	hot heading
热加工	hot working
热装	shrinkage fitting
熔融沉积成型	fused deposition modeling（FDM）
熔融混炼	melt mixing
柔性模	flexible die
柔性制造单元	flexible manufacturing cells（FMC）
柔性制造技术	flexible manufacturing technology（FMT）

柔性制造系统	flexible manufacturing system (FMS)
3D(三维)打印	three dimensional printing
闪光焊	flash welding
上料	loading
设备负荷率	machine load rate
设计基准	design datum
设计评审	design review
设计任务书	design assignment
生产纲领	production program
生产过程	production process
生产批量	production batch
生产周期	production cycle
湿装配	wet assembly
时间定额	standard time
试装	trial assembly
室温硫化	room-temperature vulcanization
手工铆接	manual riveting
数据库	data base
数控	numerical control (NC)
数控机床	numerical control machine tool
数控加工	numerical control machining
死区	dead metal zone

伺服系统	servo system
塑炼	plasticate
塑性	plasticity
塑性变形	plastic deformation
缩颈	necking
弹性变形	elastic deformation
镗床	boring machine/borer
镗孔/镗削	boring
特种加工	non–traditional machining（NTM）
填充/密封材料	packing material
填料	filler
贴胶	skim coating
铁砧/砧座	anvil
通用工装	universal tooling
涂层刀具	coating tool
推孔	hole push broaching
推削	push broaching
退刀	tool retracting
脱硫	devulcanization
脱模剂	release agent
脱膜	die releasing/stripping
外圆磨	external cylindrical grinding

弯曲	bending
往复次数	number of strokes
微细加工	micromachining
温锻	warm forging
卧式硫化机	horizontal vulcanizer
无心磨	centreless grinding (CG)
无心外圆磨床	centreless external cylindrical grinding machine
物联网	Internet of Things (IoT)
物料需求计划	material requirement planning (MRP)
铣床	milling machine/miller
铣刀	milling cutters
铣孔	hole milling
铣削	milling
下料	cropping
线焊	seam welding
卸料	unloading
行程	stroke/travel
修边	deflashing
虚拟制造	virtual manufacturing
虚拟装配	virtual assembly
悬臂	overhanging rail

旋压	spinning
选择性激光烧结	selective laser sintering（SLS）
压痕	indentation
压力加工	pressworking
压铆	press/squeeze riveting
压窝	dimple
压延	calendering
压延机	calender
压装	press fitting
延性	ductility
研孔	hole lapping
研磨	lapping
阳模	male die
摇臂	arm
液压成型	hydraulic moulding
液压夹具	hydraulic jigs/fixtures
已加工表面	machined surface
阴模	female die
应变时效硬化	strain age – hardening
永久变形	permanent deformation
油封	oil sealing
有限元分析	finite element analysis（FEA）

有限元建模	finite element modeling
预成形	preforming
预硫化	precure/prevulcanization
原始设备制造商	original equipment manufacturer（OEM）
再结晶	recrystallization
在制品	work – in – process
錾子	chisel
增材制造	additive manufacturing（AM）
轧制/滚压	rolling
粘接	gluing
展性	malleability
找正	aligning
折断	break – off
真空沉积	vacuum deposition
振动切削	vibrocutting
整修/修边	shaving/trimming
正硫化/最佳硫化	optimum cure
直接成型技术	direct moulding technology
直接金属激光烧结	direct metal laser sintering（DMLS）
制成品	finished goods
中空	hollow
主轴转速	spindle speed

注塑	injection moulding
铸件/铸造/浇注	casting
铸造	foundry
装配(件)/组件	assembly
装配过程	assembly process
装配基准	assembly datum
装配流程图	assembly flow charts
自动化	automation
自动化仓库	automated storage/retrieval system（AS/RS）
自动化加工单元	automated machining cell
自动化生产	automated production
自动进给	automatic feed
自夹紧夹具	self-clamping jigs/fixtures
总装	general assembly
组合夹具	build up jigs/fixtures
钻床	drilling machine/driller
钻孔	hole/core drilling
钻削	drilling
作业车间	job shop

试验检测
Testing & Inspection

埃里克森杯突试验	Erichsen cupping test
安装力矩	assembly/seating torque
奥斯特	Oersted（Oe）
靶	target
白斑	white spots
扳拧力矩	wrench torque
半衰期	half life
半影	penumbra
饱和	saturation
保持能力	retention capability
保证载荷	proof load
保证载荷试验	proof load test
曝光/暴露	exposure
曝光曲线	exposure chart
暴光色牢度	colour - fastness on exposure to light
爆破强度	bursting strength
比较测量	comparative measurement
边缘效应	edge effect

便携式	portable
标称频率	nominal frequency
标记/标线	reference/bench marks
标距	gauge length
标准块	reference block
标准试块	standard test block
表观密度	apparent density
表观硬度	apparent hardness
表压力	gage pressure
表针摆动范围	full indicator movement（FIM）
波	wave
补偿线圈	compensation coil
不连续(性)	discontinuity
布氏硬度	Brinell Hardness（HB）
布纹	cloth mark
采集频率	acquisition frequency
采集时间	acquisition time
参数	parameter
残余变形	offset
残余屈服点	offset yield point
残余屈服应力	offset yield stress
残余伸长率	percentage permanent elongation

残余应力	residual stress
残余载荷	residual load
测高温	pyrometry
差热分析	differential thermal analysis (DTA)
拆卸力矩	unseating torque
缠绕试验	wrapping test
常数	constant
超声波清洗	ultrasonic cleaning
超声探伤	ultrasonic flaw detection
超声显微镜	ultrasonic microscope
成形性	formability
持久断后伸长率	percentage elongation of stress – rupture
持久强度极限	stress – rupture limit
持久缺口敏感系数	stress – rupture notch sensitivity factor
持久性/耐用性	durability
持续时间	duration
冲击试验	impact test
抽样检验	sampling inspection
初始塑性伸长率	percentage initial plastic elongation
初始应力	initial stress
穿透显示	display through
磁场	magnetic field

磁场分布	magnetic field distribution
磁场计	magnetic field meter
磁场强度	magnetic field strength
磁场指示器	magnetic field indicator
磁导率	magnetic permeability
磁粉	magnetic particle
磁粉探伤	magnetic particle inspection
磁痕	magnetic particle indication
磁化	magnetizing
磁化线圈	magnetic coil
磁通	magnetic flux
磁通穿透深度	flux penetration
磁通密度	magnetic flux density
磁悬液	suspension
磁轭	yoke
磁滞	hysteresis
刺激	stimulation
粗糙度	roughness
促进剂	accelerator
脆性/脆化	brittleness/brittlement
脆性断裂	brittle fracture/rupture
脆性断裂百分率	percent brittle fracture

脆性温度	brittleness temperature
搭接	lap joint
大气压	atmospheric pressure
单剪试验	single shear test
当量	equivalent
当量法	equivalent method
导电性/电导率	conductivity
导向件	guiding element
低温冲击脆性试验	low – temperature impact brittleness test
低温蠕变	cold creep
低温研磨	cryo – grinding
低周疲劳试验	low – cycle fatigue test
滴定溶液	titration solution
电磁辐射	electromagnetic radiation
电磁检测	electromagnetic testing
电磁耦合	electromagnetic coupling
电磁铁	electromagnet
电感耦合等离子体	inductively coupled plasma（ICP）
电极	electrode
电流磁化法	current magnetization method
调节/调整	conditioning
调节剂	conditioning agent

钉芯断裂载荷	mandrel break load
定量的	quantitative
定伸应力	tensile stress at a given elongation
定位	location
定性的	qualitative
定应力伸长率	elongation at a given stress
定影	fixing
动平衡试验	dynamic balance test
动态电流	dynamic currents
动态分析	dynamic analysis
动态检测	dynamic measurement
动态疲劳	dynamic fatigue
动态应变	dynamic strain
端部效应	end effect
断后标距	final gauge length (after fracture)
断裂	fracture
断裂韧性/韧度	fracture toughness
断裂伸长率	elongation at break/rupture
断裂总延伸率	percentage total extension at fracture
断面收缩率	reduction of area
堆积密度	bulk density
对比度	contrast

对比灵敏度	contrast sensitivity
对称性	symmetry
惰性气体	inert gas
额定压力	rated pressure
轭式线圈	yoked coil
发泡剂	blowing agent
法向力	normal force
法向应力	normal stress
反射	reflection
范围	range
防腐蚀剂	corrosion inhibitor
防护材料	protective material
放气	bumping
飞边	flash
分贝	decibel（dB）
分辨率	resolution
分量分析	component analysis
分散剂	dispersing agent
粉化	chalking
峰值电流	peak current
辐射	radiation
腐蚀	corrosion

复数剪切模量	complex shear modulus
复制	replication
覆盖长度	length of coverage
覆盖区	area of coverage
干沉积荧光渗透剂	fluorescent dry deposit penetrant
干法	dry method
干涉	interference
干显像剂	dry developer
干燥剂	desiccant
干燥箱	drying oven
感应传感器	inductive sensor
刚度/刚性	rigidity
高能 X 射线	high energy X‑rays
高温	elevated temperature
高温拉伸	elevated temperature tensile
高温双剪试验	elevated temperature double shear test
高周疲劳试验	high‑cycle fatigue test
工程应变	engineering strain
工程应力	engineering stress
工作温度	working temperature
工作压力	working pressure
公差	tolerance

公称应变	nominal strain
公称应力	nominal stress
共振曲线	resonance curve
管环扩口试验	ring expanding test of tube
光弹试验	photoelastic test
光电吸收	photoelectric absorption
光谱仪	spectrometer
光学密度计	densitometer
国际橡胶硬度	international rubber hardness degrees (IRHD)
过乳化	over emulsification
过洗	overwashing
过载恢复时间	overload recovery time
合格证	conformity certificate
横截面	cross section
横向分辨率	transverse resolution
横向应变	transversal strain
红外辐射	infrared radiation
红外检测	infrared testing
红外热成像	infrared thermal imaging
红外热像法	infrared thermography
红外吸收光谱	infrared absorption spectra

红外线	infrared ray
宏观检查	macro examination
后硫化/残余硫化	aftercure
后乳化	post emulsification
化学侵蚀	chemical attack
化学吸附	chemical absorption
化学粘接	chemical bonding
划痕硬度	scratch hardness
环境温度	ambient temperature
灰分	ash
灰雾	fog
灰雾密度	fog density
挥发性脂肪酸值	volatile fatty acid number
回弹性	resilience
回归分析	regression analysis
活度	activity
活化剂	activator
霍尔系数	Hall coefficient
霍尔效应传感器	Hall effect sensor
机电耦合系数	electro - mechanical coupling factor
机械应变	mechanical strain
积累因子	build - up factor

基准线法	reference line method
激活	activation
极限抗拉强度	ultimate tensile strength（UTS）
几何不清晰度	geometric unsharpness
几何效应	geometric effect
计量检验	quantitative inspection
计量仪器	measuring instruments
计量装置	measuring apparatus/metering device
剂量当量	dose equivalent
剂量器	dosemeter
加聚反应	addition polymerization
加速老化试验	accelerated ageing test
加速天候老化试验	accelerated weathering test
夹紧力	clamping force
假缺陷	artifact
假象	nonrelevant indication
间接曝光	indirect exposure
剪切	shear
剪切断裂百分率	percent shear fracture
剪切模量	shear modulus
剪切强度极限	Fsu
剪切应力	shear stress

检查/检验	inspection
检验/验证	verification
检验批	inspection lot
碱洗	alkaline cleaning
鉴定	identification
降解	degradation
焦距	focal distance
接触腐蚀	contact corrosion
接触污染	contact stain
结构用托板紧固件搭接接头剪切试验	structural panel fastener lap joint shear test
结块	knuckles
介电常数	dielectric constant
界面	interface/boundary
金相(学)	metallography
金相分析	metallographic analysis
金相检验	metallographic examination
金属管扩口试验	drift – expanding test of tube
金属管压扁试验	flattening test of tube
近表面缺陷	near surface defect
浸没清洗	immersion rinse
浸渍	dipping

晶粒度	grain size
径向的	radial
静摩擦	static friction
静平衡试验	static balance test
静态分析	static analysis
静态检测	static measurement
静态疲劳	static fatigue
静态应变	static strain
静载荷	static load
居里点	Curie point
居里温度	Curie temperature
卷边试验	flange test
卷曲试验	coiling test
绝对测量	absolute measurement
绝对零度（−273℃）	absolute zero
绝对湿度	absolute humidity
绝对温度	absolute temperature
绝对压力	absolute pressure
绝对值	absolute value
均方根应力	root – mean – square（rms）stress
卡尺	caliper
抗冲击性能	impact resistance

抗剪/剪切强度	shear strength
抗静电剂	antistatic agent
抗拉/拉伸强度	tensile strength
抗凝剂	anti‑coagulant
抗扭强度	torsional strength
抗压强度	compressive strength
抗氧剂	antioxidant
抗粘连剂	anti‑blocking agent
可见光	visible light
可靠性	reliability
可燃性	flammability
可溶显像剂	soluble developer
可水洗型渗透剂	water‑washable penetrant
可行性	feasibility
可行性试验	feasible test
可追溯性	traceability
空洞	void
空载试验	no‑load test
孔眼	cell
扩口试验	flaring test
扩散	diffusion
拉力试验	tension test

拉伸模量	tensile modulus
拉伸疲劳	tensile fatigue
拉伸强度极限	Ftu
拉伸试验	tensile test
拉伸应力	tensile stress
拉伸永久变形	tensile set
里氏硬度	Leeb hardness（HL）
力	force
力矩/扭矩	torque
力学/机械性能	mechanical property/behaviour
力学试验	mechanical testing
沥滤	leaching
粒度	particle size
亮度	luminance
量块	gauge block
裂纹	crack/fissure
裂纹长度	crack length
裂纹扩展量	crack extension
裂纹增量	crack growth
临界常数	critical constant
临界应变	critical/threshold strain
临界应力	critical stress

灵敏度	sensitivity
零位平衡	null balance method
流滴时间	drain time
硫化剂	vulcanizing agent
硫化速率	cure rate
硫化仪	curemeter
硫酸	sulfuric acid
露点	dew point
滤光板/滤波器/过滤器	filter
滤网	screen
洛氏硬度	Rockwell hardness（HR）
马氏硬度	Martens hardness（HM）
脉冲	pulse
脉冲载荷	pulse load
盲区	dead zone
铆接接头强度	strength of a riveted joint
铆接力	rivet setting load
铆接性	driveability
门尼粘度	Mooney viscosity
密度	density
密封	seal

密封胶	sealant
模量	modulus
摩擦	friction
摩擦系数	friction coefficient
摩尔质量	molar mass
磨耗/磨损	abrasion/wear
磨耗量	abrasion loss
N 次循环后的疲劳强度	fatigue strength at N cycles
耐光牢度	light fastness
耐磨性	abrasion resistance
耐磨指数	abrasion resistance index
挠曲试验机	flexometer
拧出/松脱力矩	breakaway torque
拧紧力矩	tightening torque
拧入力矩	entry torque
拧松力矩	untightening torque
凝固剂	coagulant
凝固剂浸渍	coagulant dipping
凝聚	coagulation
扭出	torque – out
扭矩-拉力试验	torque tension test

扭转	torsion
浓度	concentration
浓缩物	concentrates
努氏硬度	Knoop hardness（HK）
耦合	coupling
耦合剂	couplant
耦合损失	coupling losses
耦合系数	coupling factor
排布	arrangement
喷嘴	spray nozzle
膨胀	expansion
疲劳	fatigue
疲劳变形能力	fatigue deformability
疲劳极限	fatigue limit
疲劳破坏	fatigue breakdown
疲劳强度	fatigue strength
疲劳试验	fatigue test
疲劳寿命	fatigue life
疲劳延性系数	fatigue ductility coefficient
疲劳应力	fatigue stress
频率	frequency
频率常数	frequency constant

平均粒径	average particle diameter
平面应变断裂韧度	plane – strain fracture toughness
平行长度	parallel length
评定	evaluation
评定区	area of interest
破坏性试验	destructive test
气泡	blister/bubble
千分尺	micrometer
翘曲	warping
切向取样试验	tangential test
亲和力	affinity
亲水性乳化剂	hydrophilic emulsifier
亲油性乳化剂	lipophilic emulsifier
侵蚀	erosion
氢氟酸	hydrofluoric acid
清晰度	definition
清洗	rinse
区域定位	zone location
屈服点延伸率	percentage yield point extension
屈服拉伸应力	tensile stress at yield
屈服强度	yield strength
屈服伸长率	elongation at yield

屈服位移	general yield displacement
屈服应力	yield stress
去离子水	deionized water
缺口	notch
缺陷	defect
缺陷检出灵敏度	defect detection sensitivity
热成像系统	thermal imaging system
热冲击试验	thermal shock test
热导率	thermal conductivity
热电偶	thermocouple
热分辨率	thermal resolution
热降解	thermal degradation
热扩散	thermal diffusion
热扩散率	thermal diffusivity
热敏剂	heat sensitizer
热敏浸渍	heat – sensitive dipping
热像图	thermal image/thermogram
热像仪	infrared camera
热应变	thermal strain
热应力	thermal stress
人工缺陷	artificial defect
韧度/韧性	toughness

溶剂	solvent
溶解	dissolution
溶胀	swelling
溶胀度	swellability
蠕变	creep
蠕变断后伸长率	percentage elongation after creep rupture
蠕变断裂时间	creep rupture time
蠕变强度	creep strength
蠕变曲线	creep curve
蠕变伸长率	percentage creep elongation
蠕变伸长时间	creep elongation time
蠕变试验	creep test
蠕变指数	creep index
乳化	emulsification
乳化剂	emulsifier
乳化渗透剂	emulsifiable penetrant
乳化时间	emulsification time
乳液聚合	emulsion polymerization
入厂检验	incoming inspection
润滑剂	lubricant
塞尺	thickness/plug gage
扫描	sweep/scan

扫描电镜	scanning electron microscope (SEM)
扫描范围	sweep range
扫描速度	sweep speed
色散	dispersion
色散介质	dispersion medium
闪点	flash point
上屈服强度	upper yield strength
邵氏硬度	Shore hardness
伸长(率)	elongation
伸长计/引伸计	extensometer
伸长率	percentage elongation
渗出	bleed out
渗漏/渗透	permeation
渗透	effusion
渗透探伤	penetrant flaw detection
渗透系数	permeability coefficient
渗透性	permeability
剩磁	residual magnetism
剩磁法	residual magnetic method
失效分析	failure analysis
湿斑	wet spots
湿度	humidity

湿化学法	wet chemistry method
石墨炉	graphite furnace
时间同步显示	time synchronous display
蚀刻/侵蚀/腐蚀	etch
使用寿命	service life
势能	potential energy
试块	test block
试样	specimen
试样/试件	test specimen/piece
试样制备	sample preparation
室温	room temperature
收缩	shrinkage
手工检测	manual testing
寿命试验	life test
双剪	double shear
水膜残迹	water break
水平定位	horizontal location
水平极限	horizontal limit
水平线性	horizontal linearity
水洗	aqueous wash
撕裂强度	tear strength
塑性	plasticity

塑性断裂百分率	percent ductile fracture
塑性伸长时间	plastic elongation time
塑性应变	plastic strain
塑性应变比	plastic strain ratio
酸洗	pickling
锁紧/收口	locking
弹性	elasticity
弹性变形	elastic deformation
弹性极限	elastic limit
弹性剪切模量	elastic shear modulus
弹性介质	elastic medium
弹性模量	modulus of elasticity
弹性应变	elastic strain
探头/探针	probe
特性因素	characteristic factor
体积密度	volume density
添加剂	additive
填充剂	extender
凸耳试验	earing test
图像对比度	image contrast
图像放大	image magnification
图像增强	image enhancement

图像质量	image quality
推出	push – out
退磁	demagnetization
U 型缺口	U – notch
V 型缺口	V – notch
外观检查	visual examination
弯曲/挠曲	flexure
弯曲半径	bend radius
弯曲试验	bend test
万能拉伸试验机	universal tensile testing machine
网纹	reticulation
微分滤波器	differential filter
微观结构	microstructure
微压痕硬度	micro indentation harness
维氏硬度	Vickers hardness (HV)
位移特征值	displacement characteristic value
温度标定	temperature calibration
温度均匀性	temperature uniformity
稳定剂	stabilizer
稳定性	stability
涡流	eddy current
涡流检测	eddy current testing

污染	stain
污染物	contaminant
无机酸	mineral acid
无损检测	non – destructive testing（NDT）
物理吸附	physical absorption
X 射线探伤装置	X – ray detection apparatus
X 射线荧光	X – ray fluorescence（XRF）
吸收	absorption
吸收剂量	absorbed dose
吸收能量	absorbed energy
下屈服强度	lower yield strength
夏比摆锤冲击试验	Charpy pendulum impact test
夏比缺口冲击试验	Charpy notch impact test
显示区域	display area
显微腐蚀	micro – etching
显像时间	developing time
显影	development
显影过度	over development
线材缠绕试验	wrapping test of wire
线材反向扭转试验	reverse torsion test of wire
线材扭转试验	torsion test of wire
线圈匝数	coil turns

线性衰减系数	linear attenuation coefficient
线性应变	linear strain
相	phase
相变	phase transition
相关系数	correlation coefficient
相似定律	law of similarity
相位分析	phase analysis
橡胶屑	buffings
硝酸	nitric acid
校准/标定	calibrate
校准试块	calibration block
楔垫拉伸	wedge tensile
楔负载拉力试验	tensile test under wedge loading
楔形块试验	wedge testing
谐波分析	harmonic analysis
泄漏检测	leak testing
型式试验	type test
循环	cycle
压痕	indentation/pressure mark
压痕模量	indentation modulus
压痕硬度	indentation hardness
压力差	pressure difference

压缩空气干燥	compressed air drying
压缩模量	compression modulus
压缩弹性模量	compressive modulus of elasticity
压缩应变	compression strain
压缩应力	compression stress
压缩应力松弛	compression stress relaxation
压缩永久变形	compression set
压头	indenter
延伸	extension
延伸率	percentage extension
延时扫描	delayed sweep
延性	ductility
盐雾试验	salt spray test
盐浴	salt bath
颜料	pigment
颜色/比色指数	colour index
验收	acceptance
阳极	anode
氧化皮	scale
液膜显像剂	liquid film developer
乙酸乙酯	ethyl acetate
抑制剂	inhibitor

阴极	cathode
应变	strain
应变幅	strain amplitude
应变速率	strain rate
应变硬化指数	strain hardening exponent
应力	stress
应力持久	stress durability
应力断裂	stress rupture
应力幅	stress amplitude
应力面积	stress area
应力强度因子	stress intensity factor
应力屈服	stress yield
应力松弛	stress relaxation
应力松弛曲线	stress relaxation curve
应力松弛速率	stress relaxation rate
应力速率	stress rate
应力-应变曲线	stress – strain curve
荧光	fluorescence
荧光磁粉	fluorescent magnetic powder
荧光屏	fluorescent screen
荧光渗透剂	fluorescent penetrant
荧光探伤	fluorescent inspection

硬度	hardness
硬度计	durometer
硬化	hardening
硬化剂	stiffener
永久变形	permanent set
游标卡尺	vernier caliper
游离硫	free sulfur
有机酸	organic acid
有效磁导率	effective magnetic permeability
有效渗透深度	effective depth of penetration
诱导期	induction time
预清洗	pre – cleaning
预应变	pre – strain
预应力	pre – stress
预载	preload
元素分析	elemental analysis
元素符号	atomic symbols
原始标距	original gauge length
原子发射光谱法	atomic emission spectrometry（AES）
原子吸收光谱法	atomic absorption spectrometry（AAS）
运动粘度/动力粘度	kinematic viscosity
杂质	impurity

载荷-位移曲线	load – displacement curve
载液	carried fluid
增塑剂	plasticizer
粘度	viscosity
粘合/粘着	adhesion
粘合促进剂	adhesion promoter
粘合强度	adhesion strength
粘合系数	adhesion factor
粘接剂	adhesive/bonding agent
粘连	blocking
遮蔽	masking
遮挡介质	blocking/masking medium
折射	refraction
真空暗盒	vacuum cassette
真应变	true strain
真应力	true stress
振动	vibration
振动试验	vibration test
振幅	amplitude
蒸发/汽化	evaporation
蒸馏水	distilled water
蒸汽除油	vapor degreasing

直接接触法	direct contact method
直接曝光成像	direct exposure imaging
直流等离子体	direct current plasma（DCP）
指示	indication
指向性	directivity
质量衰减系数	mass attenuation coefficient
质谱分析法	mass spectrometry
质谱仪	mass spectrometer（MS）
滞后回线	hysteresis loop
置信区间	confidence interval
周向磁场	circumferential field
轴向拉伸	axial tensile
轴向应变	axial strain
轴向应力	axial stress
轴向载荷	axial load
主动式热像检测	active thermal - graphic testing
注射量	shot
贮存期限/寿命	shelf/storage life
转变温度	transition temperature
浊度	turbidity
着色剂	colourant
着色强度	tinting strength

着色渗透剂	dye penetrant
资源共享	resource sharing
紫外线	ultraviolet ray
自锁力矩	self – locking torque
纵向分辨率	longitudinal resolution
族	family
阻尼/衰减系数	damping coefficient
阻尼常数	damping constant
阻尼块	damping block
最大锁紧力矩	maximum locking torque
作用力	active force
作用区	zone of interaction
坐标网格法	coordinate grid method
γ 射线	γ ray

材料与热处理
Material & Heat Treatment

奥氏体	austenite
奥氏体化	austenitizing
奥氏体镍铬不锈钢	austenitic Ni – Cr stainless steel
奥氏体稳定化处理	austenite stabilization
板材/厚板(料)	plate
半镇静钢/半脱氧钢	balanced steel
伴生变形	concurrent deformation
棒材/棒料	bar
包铝	clad aluminum
饱和度	degree of saturation
保护气氛	protective atmosphere
保温	holding/soaking
贝氏体	bainite
比强度	specific strength/strength – to – weight ratio
比重	specific gravity
变形	deformation
变形量	amount of deformation
变形孪晶	deformation twin

变形织构	deformation texture
标准混炼胶	standard compound
表面淬火	surface hardening
表面热处理	surface heat treatment
表面渗碳层	carburized case
表面退火	skin annealing
表面脱碳层	carbon – free surface layer
表面硬化	case hardening
玻璃钢	glass reinforced plastics (GRP)
泊松比	Poisson's ratio
薄板(料)/板料	sheet
不完全退火	incomplete/partial annealing
不锈钢	stainless steel
步进式炉	walking beam furnace
残余应力	residual/internal stress
层状断裂/断口	laminar fracture
差温加热	differential heating
常规成形工艺	conventional forming technique
超高频感应加热装置	ultra high frequency induction heating equipment
超高强度钢	extra – high tensile steel
超临界状态	above – critical state

沉淀硬化不锈钢	precipitation hardening stainless steel
成分一致性	consistency of composition
冲击韧性试验	impact toughness test
冲压凹痕	cupping/dishing
臭氧老化	ozone ageing
初生 α 相	primary alpha
除鳞/去氧化皮	descale
吹炼炉	converting furnace
穿晶断裂	transgranular cracking
传热系数	heat transfer coefficient（HTC）
传送带式炉	conveyor belt furnace
粗晶粒钢	coarse - grained steel
脆性断裂源	brittle fracture initiation
脆性腐蚀特性	brittle erosion behaviour
脆性-塑性转变温度	brittle - ductile transition temperature
淬火	quenching/hardening
淬火槽	quenching tank
淬火介质	quenchant
淬火硬度	as - quenched hardness
淬透性	hardenability
淬硬层深度	depth of hardening zone
淬硬性	hardening capacity

带料	strip
带状织构	zonal texture
带状组织	banded/zonal structure
单相(的)	uniphase
单向应力	uniaxial stress
单轴的	uniaxial
氮	nitrogen (N)
等温淬火	isothermal hardening
等温正火	isothermal normalizing
等温转变	isothermal transformation
等轴晶区/晶带	zone of equiaxed crystal
等轴组织	equiaxed structure
低合金钢	low – alloy steel
低温回火	low temperature tempering
滴注式气氛	drop feed atmosphere
底开式炉	drop bottom furnace
电磁搅拌	electromagnetic agitation
电弧炉	Electric Arc Furnace (EAF)
电炉	electric furnace
电渣重熔	Electroslag Remelting (ESR)
电子束炉	Electron Beam Furnace (EBF)
电阻加热	resistance heating

电阻炉	resistance furnace
调节/调整	conditioning
调质钢	quenched/hardened and tempered steel
丁苯橡胶	styrene - butadiene rubber
丁腈橡胶	nitrile rubber
端淬试验	Jominy end quenching test
断口	fracture
断口金相照片	fractograph
断口晶粒度	fracture grain size
断面形状	form of section
堆积胶	bank
多工区炉	multi - working zone furnace
多晶的	multi - crystal
多孔柔性聚合材料	cellular polymeric flexible material
多相（的）	multiphase
多元高速钢	complex high - speed steel
多元共渗剂	multicomponent diffusion agent
惰性气氛	inert atmosphere
惰性填料	inert filler
二次退火	second annealing
二相合金	duplex alloy
二元相图	binary phase diagram

发蓝退火	blue annealing
发裂	shatter crack
钒	vanadium（V）
反向应力	back stress
芳烃油	aromatic oil
放热式气氛	exothermic atmosphere
非金属夹杂	non – metallic inclusion
废料	scrap
沸腾钢	boiling steel
分级淬火	broken/graded hardening
分批退火	batch annealing
酚醛树脂	phenolic resin
粉末冶金	powder metallurgy
风冷装置	forced air cooling device
氟烃橡胶	fluorocarbon rubber
辅助电极	auxiliary electrode
负/反偏析	negative segregation
复合材料	composite material
复合应力	combined stress
副产品	by – product
覆皮	applied skin
γ 相变	γ transformation

坩埚	crucible
感应回火	induction tempering
感应加热表面淬火	induced surface hardening
感应退火	induction annealing
感应线圈	induction coil
刚度	rigidity
高合金钢	high alloy steel
高密度夹杂	high density inclusion (HDI)
高能束热处理	high energy beam heat treatment
高频感应加热装置	high frequency induction heating equipment
高强度低合金钢	high strength low alloy steels (HLS)
高强度金属	high-duty metal (HDM)
高速钢	high speed steel (HSS)
高碳马氏体	high carbon martensite
高温回火	high temperature tempering
高温计	pyrometer
高温氧化	high temperature oxidation (HTO)
膏化	creaming
镉	cadmium (Cd)
各向同性	isotropy
各向异性	anisotropy/aeolotropism
各向异性材料	anisotropic material

铬	chromium (Cr)
铬酸盐底层涂料	chromate primer
工序间退火	inter – process annealing
工业纯钛	commercially pure titanium
工业金属	commercial metal
工艺材料	technological materials
汞	mercury (Hg)
共晶-包晶反应	eutectic – peritectic reaction
共晶组织	eutectic structure
共析钢	eutectoid steel
钴	cobalt (Co)
固化炉	curing furnace
固溶	solution
固溶热处理	solution heat treatment
固溶退火	solution annealing
固溶硬化	solution hardening
刮痕	scratch
拐点/回折点	inflexion point
管材	tubing
光洁度	degree of finish
光亮淬火	bright quenching/hardening
光亮热处理	bright heat treatment

光亮退火	bright annealing
光谱分析	analysis of spectra
龟裂	crazing
硅	silicon (Si)
贵金属	noble metal
辊底式炉	roller hearth furnace
国际钢铁学会	International Iron and Steel Institute (IISI)
国际焊接学会	International Institute of Welding (IIW)
过共析钢	hypereutectoid steel
过冷度	degree of undercooling
过热	overheating
过烧	burning
合成橡胶	synthetic rubber
合金成分偏析	alloy segregation
合金钢	alloy steel
合金工具钢	alloy tool steel
合金时效	alloy ageing
黑斑/黑点	black dot
黑色金属	ferrous metal
恒温箱	attemperator
恒压、恒容和恒温	constant pressure, volume and temperature (p. v. t.)

横向/纵向磁通加热	transverse/longitudinal flux heating
红外炉	infra – red furnace
宏观组织	macrostructure
化学抛光	chemical brightening
化学侵蚀	chemical attack
化学清洗	chemical cleaning
化学热处理	thermo – chemical treatment
还原剂	reducing agent
还原气氛	reducing atmosphere
还原态金属	as – reduced metal
环氧树脂	epoxy resin
环氧橡胶	epoxide rubber
黄铜	brass
回火	tempering
回火脆性	temper brittleness
回火时效	temper ageing
混炼胶	compound
活化处理	activating treatment
活性气氛	active atmosphere
活性碳	activated carbon
积蓄热	accumulated heat
基体	matrix

基体沉淀/析出	matrix precipitation
基体金属	basis metal
激光加热	laser heating
挤压型材	extruded shapes
加热介质	heating medium
加热室/炉膛	heating chamber
加热制度	heating schedule
加速腐蚀	accelerated corrosion
加速冷却	accelerated cooling
加速燃烧	accelerated combustion
加速蠕变	accelerated creep
加速时效	accelerated ageing
夹具	fixture
夹杂物	inclusion
间接电阻加热	indirect resistance heating
间歇式炉	batch type furnace
间歇式炉/非连续式炉	discontinuous furnace
碱金属	alkali metal
碱浴炉	alkali bath furnace
交货状态	as - received condition
焦烧	scorch

矫直/校直机	straightening machine
结构钢	structural steel
结合硫	combined sulfur
结合橡胶	bound rubber
结晶	crystallization
金相检验	metallographic examination
金属工艺学	technology of metals
金属基复合材料	metal – matrix composites（MMCs）
金属间化合物	intermetallic compound
晶胞	cell
晶格	lattice/grate
晶格/点阵缺陷	lattice defect
晶格膨胀	lattice dilatation
晶格异质	lattice heterogeneity
晶核中心	germ nucleus
晶间腐蚀	intergranular attack（IGA）
晶间偏析	intercrystalline segregation
晶间收缩裂纹	intercrystalline shrinkage crack
晶间氧化	intergranular oxidation（IGO）
晶界	grain boundary
晶粒	grain
晶粒度	grain size

晶粒号	grain size number
晶粒流线	grain flow
晶粒细化处理	grain – refinement treatment
晶轴常数	axial constants
精炼钢	refined steel
精炼炉	affinage furnace
井式炉	pit furnace
局部热处理	local heat treatment
局部渗碳	localized/selective carburizing
局部退火	selective annealing
聚合	polymerization
聚合材料	polymeric material
聚合物	polymer
聚氯乙烯	chlorinated polyethylene/polyvinyl chloride (PVC)
聚四氟乙烯	polytetrafluoroethylene (PTFE)
聚乙烯	polyethylene (PE)
卷料	coil
绝缘材料	insulating material
均匀腐蚀	uniform corrosion
均匀加热	soak heating
均匀性/一致性	uniformity

均质钢	homogeneous steel
可控气氛热处理	controlled atmosphere heat treatment
空冷	air cooling
空气淬硬钢	air – hardening steel
空气过剩系数	air excess coefficient
空心多孔材料	cored cellular material
孔隙度	amount of porosity
快速退火	short annealing
扩散层	diffusion layer
扩散退火	diffusion annealing
扩散硬化	diffusion hardening
莱氏体	ledeburite
冷拔/冷拉	cold drawing
冷壁/内热式真空电阻炉	cold wall vacuum resistance furnace
冷变形	cold deformation
冷成形性	cold formability
冷脆性/低温脆性	cold brittleness
冷加工量	amount of cold work
冷加工硬化	cold work hardening
冷阱	condensing collector
冷卷	cold coiling

冷却室	cooling chamber
冷却制度	cooling schedule
冷时效/常温时效	cold ageing
冷塑性变形量	amount of cold plastic deformation
冷态流变	cold flow
冷轧带材	flat cold rolled strip
冷轧钢板	cold-rolled steel sheet
立方晶系	cubic system
沥青	asphalt
沥青橡胶	asphalt rubber
连续淬火回火作业线	continuous quench and temper line
连续镀锌作业线	continuous galvanizing line (CGL)
连续加工处理作业线	continuous processing line (CPL)
连续拉拔	continuous drawing
连续冷却曲线	continuous cooling curve
连续式炉	continuous furnace
连续退火	continuous annealing
链条输送式炉	chain conveyor furnace
料筐	charging basket
料盘	charging tray

裂变	fission
裂纹敏感性组成（成分）	cracking sensitivity composition
临界淬火速率	critical rate of hardening
临界点/值	threshold
临界切应变	critical shear strain
磷	phosphorus（P）
零点校验	check for zero
流痕	flow marks
硫	sulfur（S）
硫化橡胶	vulcanized rubber
六角钢	hexagonal bar
露点	dew point
炉底	hearth
炉底板	hearth plate
炉底热强度	thermo - intensity of hearth
炉顶	arch
炉盖	furnace lid
炉架	furnace frame
炉壳	furnace casing
炉冷	furnace cooling
炉料	charge

炉料转移系统	charge transfer system
炉门	furnace door
炉墙	furnace wall
炉室	furnace chamber
炉体	furnace body
炉温均匀度	furnace temperature uniformity
炉温稳定度	furnace temperature stability
铝	aluminium (Al)
氯丁橡胶	chloroprene rubber
孪晶	twin
孪晶边界	twin boundary
络合物	complex compound
麻点/痘痕	pockmark
马氏体	martensite
马氏体时效钢	maraging steel
脉冲感应加热装置	pulse induction heating equipment
毛坯/胶坯	blank
美国钢铁学会	American Iron and Steel Institute (AISI)
美国金属学会	American Society of Metals (ASM)
镁	magnesium (Mg)
锰	manganese (Mn)
弥散相硬化	dispersed phase hardening

面心立方（晶格）	face – centered cubic (FCC)
模锻/模压	contour forging
母炼胶	master batch
母料	mother stock
母相	parent phase
钼	molybdenum (Mo)
内热式浴炉	internally – heated bath furnace
耐高热镍基合金	Waspaloy
耐候老化试验/户外暴露试验	exposure test
耐碱钢	alkali proof steel
耐磨钢	abrasion resistant steel
挠度	amount of deflection
能带理论	band theory
铌	niobium/columbium (Nb)
镍	nickel (Ni)
凝胶	gel
牌号	grade
抛光剂/增亮剂	brightener
配方	formulation
喷砂清理	abrasive blast cleaning
喷霜	bloom

喷丸硬化	ball peening
硼	boron (B)
批次	batch
批号	batch number
疲劳	fatigue
偏析	segregation
频率和严重度	frequency&severity (F/S)
平均粒度	average grading
平均温差	average temperature difference
气淬	air quench
气淬真空电阻炉	gas – quenching vacuum resistance furnace
气候老化/风化	weathering
气体淬火介质	gas quenching medium
气体净化装置	gas purification equipment
牵引式炉	drawing furnace
铅	lead (Pb)
前室	front vestibule
青铜	bronze
氢	hydrogen (H)
氢脆	hydrogen embrittlement
清洗设备	rinsing equipment
清洗作业线	cleaning line

球化退火	spheroidizing annealing
屈氏体	troostite
去应力退火	stress relief annealing/stress relieving
全脱氧钢/全镇静钢	dead steel
缺胶	bareness
缺口弯曲试验	nick bend test
缺陷	defect
燃烧室	combustion chamber
热壁/外热式真空电阻炉	hot wall vacuum resistance furnace
热变形温度	heat distortion temperature（HDT）
热处理	heat treatment
热处理炉	heat treatment furnace
热脆材料	hot short material
热脆开裂	hot short cracking
热脆区	hot shortness zone
热当量	heat equivalent
热浸镀锌浴	zinc dipping baths
热平衡	balance of heat
热塑性橡胶	thermoplastic rubber
热稳定性	thermostability
热应力	thermal stress

热影响区	heat affected area
热浴淬火	hot bath hardening
人工气候老化	artificial weathering
人工时效	artificial ageing
韧化(处理)	toughening
韧性/韧度	toughness
韧性断口	tough fracture
日本金属学会	Japanese Institute of Metals (JIM)
熔化/(核)聚变	fusion
柔度	flexibility
柔量	compliance
蠕变	creep
色谱分析	chromatographic analysis
烧嘴	burner
渗氮	nitriding
渗氮气氛	nitriding atmosphere
渗碳	carburizing
升温时间	heating up time
生胶	raw rubber
生热/热积累	heat build-up
失效	failure
石墨	graphite

石墨化退火	graphitizing treatment
时效/老化	ageing
时效稳定性	ageing stability
时效硬化沉淀	age hardening precipitation
时效硬化钢	age hardening steel
收缩	contraction
收缩量	amount of contraction
收缩裂纹	check crack
树脂	resin
双金属制品	bimetallic article
双联淬火槽	duplex quenching tank
双频感应加热装置	double-frequency induction heating equipment
双相奥氏体铁素体不锈钢	duplex austenitic-ferritic stainless steel
双相钢	dual phase steel
双液淬火槽	dual-liquid quenching tank
双重退火	duplex annealing
水淬	water quenching
水韧处理	water toughening
瞬时应力	instantaneous stress
塑料	plastics

塑性变形区	zone of plastic deformation
酸洗	pickling
缩孔	contraction cavity
钛	titanium（Ti）
钛合金	titanium alloy
炭黑	carbon black/soot
炭黑焦烧	black scorch
碳	carbon（C）
碳含量	carbon content
碳含量分布	carbon profile
碳化物/硬质合金	carbide
碳素工具钢	carbon tool steel
碳纤维	carbon fibre
陶瓷基复合材料	ceramic - matrix composites（CMCs）
特性冷却曲线	characteristic cooling curve
特优钢	extra - fine steel
体积弹性模量	bulk modulus
体心立方（晶格）	body centred cubic（BCC）
天然橡胶	natural rubber
条料/杆	rod
铁	iron（Fe）
铁素体	ferrite

铁素体不锈钢	ferritic stainless steel
铁素体基体	ferritic matrix
同位素	isotope
同相	homophase
铜	copper（Cu）
涂覆织物	coated fabric
推送式炉	pusher furnace
退氮	denitriding
退火	annealing
退火脆性	annealing brittleness
退火工序	annealing operation
退火炉	annealing furnace
退火酸洗线	annealing – pickling line
退火周期/制度	annealing cycle
脱氢处理	hydrogen relief treatment
脱碳	decarburization
弯曲蠕变试验	bending creep test
完全退火	full annealing
微空洞/微孔	microvoid
微缩孔	microshrinkage
位错	dislocation
位错受阻	hindering of dislocation

温差应力	temperature stress
稳定(化)钢	stabilized steel
稳定化处理	stabilizing treatment
稳定化退火	stabilizing annealing
稳定性	stability
钨	tungsten（W）
无机盐水溶液淬火介质	inorganic salt solution quenching medium
无碳不锈钢	carbon – free stainless steel
无碳铁素体	carbon – free ferrite
吸热式气氛	endothermic atmosphere
析出/沉淀	precipitation
硒	selenium（Se）
稀土金属	rare earth metal
稀有元素	rare element
锡	tin（Sn）
细晶粒钢	fine grained steel
显微组织	microstructure
线材/丝材	wire
相	phrase
相变脆性	transformation embrittlement
相变潜热	intent heat of phase change

相变退火	transformation annealing
相变温度	temperature of phase change
相变应力	transformation stress
相对湿度	relative humidity
相界面	interphase boundary
箱式淬火炉	sealed box type quenching furnace
箱式热空气老化	air oven ageing
橡胶	rubber
芯棒	cored bar
锌	zinc（Zn）
形变带	bands of deformation
形变热处理	thermo - mechanical treatment（TMT）
形状记忆合金	shape memory effect（SME）alloy
旋转淬火	rolling quenching
循环冷却系统	circulation cooling system
压痕面积	area of indentation
压升率	pressure rising rate
亚共析钢	hypoeutectoid steel
亚晶界	subboundary
亚晶粒/次级晶粒	sub - grain
亚临界奥氏体	subcritical austenite
亚组织	substructure

氩氧脱碳	Argon – Oxygen Decarburization（AOD）
烟道	flue
延展度	extensibility
盐浴炉	salt bath furnace
验收检验	acceptance inspection
阳极	anode
阳离子	cation
杨氏模量	Young's modulus
氧	oxygen（O）
氧化	oxidation
氧化发黑处理	black oxide finish
氧化气氛	oxidizing atmosphere
氧压老化	oxygen – pressure ageing
液态床渗碳	fluidized bed carburizing
钇	yttrium（Y）
乙丙橡胶	ethylene propylene rubber（EPR）
异常晶粒长大	abnormal grain growth
易切削钢	free – machining steel
银	silver（Ag）
应变时效钢	strain aged steel
应变硬化	strain – hardening
应力腐蚀穿晶断裂	transgranular stress corrosion cracking

英国钢铁协会	British Iron and Steel Institute（BISI）
硬质合金	hard alloy
优质结构钢	high quality structural steel
优质碳素结构钢	carbon constructional quality steel
油浴炉	oil bath furnace
游离铁素体	free ferrite
有色金属	non – ferrous metal
浴炉	bath furnace
预热	preheating
元素符号	atomic symbols
原材料	raw material
原胶	virgin rubber
杂质	impurity
载气	carrier gas
再结晶	recrystallize
再结晶退火	recrystallization annealing
再生相/次生相	minor phase
再现性	reproducibility
轧制状态	as – rolled condition
粘度	viscosity
罩式炉	bell furnace
针状马氏体	acicular martensite

真空	vacuum
真空淬火	vacuum hardening/quenching
真空电弧重熔	vacuum arc remelting (VAR)
真空电阻炉	vacuum resistance furnace
真空度	amount of vacuum
真空回火	vacuum tempering
真空离子轰击热处理炉	ion - bombarding heat treatment vacuum furnace
真空离子渗碳炉	ion - carburizing vacuum furnace
真空炉	vacuum furnace
真空热处理	vacuum heat treatment
真空退火	vacuum annealing
振底式炉	shaker hearth furnace
整体热处理	bulk heat treatment
正火	normalizing
正挤压	forward extrusion
正温度系数	positive temperature coefficient (PTC)
织构/纹理	texture
中间合金	master alloy
中间退火	interstage annealing
中频感应加热装置	medium frequency induction heating equipment

中温回火	average tempering/medium temperature tempering
中性气氛	neutral atmosphere
中性盐	neutral salt
重复性	repeatability
重金属	heavy metals（H. M.）
重力输送式炉	gravity feed furnace
轴对称变形	axisymmetric deformation
轴对称挤压	axisymmetric extrusion
轴对称拉伸	axisymmetric drawing
珠光体	pearlite/pearlyte
贮存老化	shelf ageing
柱状晶区/晶带	zone of columnar crystal
铸态金属	as‑cast metal
铸造合金	casting alloy
转底式炉	rotary hearth furnace
自然气氛	natural atmosphere
自然时效/老化	natural ageing
组织/结构	structure
最佳取向	preferred orientation
最终热处理	final heat treatment

表面处理
Surface Treatment

白雾	white haze
比色法	colorimetry
比重杯	specific gravity cup
标准氢电极	standard hydrogen electrode
表面不连续	surface discontinuity
表面处理	finish/surface treatment
表面粗糙度	surface roughness
表面活性剂	surface active agent/surfactant
表面结构	surface texture
表面孔隙率	apparent porosity
表面污染	surface contamination
波纹度	waviness
剥离	spalling
剥离/去皮/起皮	peeling
剥离强度	peeling strength
剥落	flaking
不连续水膜	water break
布氏（B 型）粘度计	Brookfield viscometer

残余应力	residual stress
草酸阳极化	oxalic anodic oxidation
侧抽风喷漆室	spray booth of side exhaust
超声波清洗	ultrasonic cleaning
沉积速率	deposition rate
除尘装置	dust collector
除锈	descaling/derusting/rust removal
除油	degreasing
粗糙度	roughness
粗化	coarsening/roughening
脆性	brittlement/brittleness
打底	strike/underplating
大气暴露试验	atmospheric exposure test
大气腐蚀	atmospheric corrosion
等离子喷涂	plasma spraying
点滴腐蚀试验	dropping corrosion test
点蚀	pitting/spot corrosion
电(解)抛光	electropolishing
电导率	conductivity
电镀	electroplating
电化当量	electrochemical equivalent
电化学	electrochemistry

电化学保护	electrochemical protection
电化学处理	electrochemical treatment
电化学腐蚀	electrochemical corrosion
电化学极化/活化极化	activation polarization
电化学抛光	electrochemical polishing
电极	electrode
电解	electrolyze
电解除油	electrolytic degreasing
电解清洗	electrolytic cleaning
电解酸洗	electrolytic pickling
电解液	electrolytic solution
电解质	electrolyte
电离度	degree of ionization
电流密度	current density
电流效率	current efficiency
电偶腐蚀	galvanic corrosion
电位-电流密度曲线	potential – current density curve
电泳沉积	electrophoretic deposition
吊具	sling
丁酮	methyl ethyl ketone

毒性	toxicity
镀后处理	postplating
钝化	passivation
多层电镀	multilayer plating
惰性阳极	inert anode
二硫化钼	molybdenum disulfide（MoS_2）
二硫化钼润滑剂	molykote
防蚀	corrosion protection
防锈油	rust preventive oil
分层	split layer
分散能力	throwing power
封闭	sealing
缝隙腐蚀	crevice corrosion
辐射对流烘干室	joint radiation convection oven
辅助电极	auxiliary electrode
腐蚀	corrosion
腐蚀电池	corrosion cell
腐蚀电流	corrosion current
腐蚀电位/电势	corrosion potential
腐蚀深度	corrosion depth
负极	negative electrode
复合电镀	composite plating

富锌底漆	zinc rich primer
干膜润滑剂	solid film lubricant
干磨	dry grinding
干式喷漆室	dry spray booth
干燥	drying
干燥箱	drying oven
隔膜	diaphragm
铬酸盐处理/铬化	chromating
铬酸阳极化	chromate anodic oxidation
固化	curing
固化剂	firming/solidified agent
刮伤	scratch
挂镀	rack plating
光亮电镀	bright plating
光泽	luster
光泽度	gloss level
龟裂/微裂	crazing
辊涂	roll coating
滚镀	barrel plating
滚光	barrel burnishing
哈林槽	Haring cell
海洋腐蚀	marine corrosion

恒应变应力腐蚀破裂试验	constant strain SCC test
恒载荷应力腐蚀破裂试验	constant load SCC test
烘干	baking
化学镀	chemical plating
化学钝化	chemical passivation
化学腐蚀	chemical corrosion
化学抛光	chemical polishing
化学气相沉积	chemical vapor deposition (CVD)
化学清洗	chemical cleaning
化学氧化	chemical oxidation
化学转化膜	chemical conversion coating
环氧底漆	epoxy primer
活度	activity
活化	activation
活化剂	activator
霍尔槽	Hull cell
机械打磨	mechanical sanding
机械镀	mechanical plating
机械抛光	mechanical polishing
机械清洗	mechanical cleaning

基体材料	basis material
基体材料/底材	substrate
激光电镀	laser electroplating
激光溶覆	laser cladding
激光釉化	laser glazing
极化	polarization
极间距	interelectrode distance
夹杂等级	inclusion rating
碱脆	caustic embrittlement
碱洗	alkaline cleaning
搅拌装置	stirring equipment
接触腐蚀	contact corrosion
界面张力	interfacial tension
金属变色	tarnish
浸镀	immersion plate
浸亮	bright dipping
浸涂	dip coating
晶间腐蚀	intercrystalline corrosion/intergranular attack/intergranular corrosion
晶粒度	grain size
静电喷涂	electrostatic spraying
局部腐蚀	local corrosion

局部修补	spot repair
聚氨酯面漆	polyurethane topcoat
绝缘层	insulated layer
均匀腐蚀	uniform corrosion
抗高温氧化试验	test of resistance to high temperature oxidation
孔隙率	porosity
库仑计	coulombmeter
库仑效率	coulomb efficiency
扩散层	diffusion layer
扩散控制	diffusion control
离心干燥机	centrifuge
离子镀	ion plating
离子气相沉积	ion vapor deposition (IVD)
离子注入	ion implantation
凉干(流平)	flash off
临界电流密度	critical current density
磷酸盐/磷化处理	phosphating
硫酸阳极化	sulphoacid anodic oxidation
露底	lack of coverage
麻点/坑	pit
脉冲电镀	pulse plating

脉冲阳极化	pulse anodizing
面漆	finish/top coat
敏化	sensitization
模拟腐蚀试验	simulative corrosion test
膜厚测量仪	film thickness gauge
磨光	grinding
磨料	abrasive
内应力	internal stress
耐热性	heat resistance
耐蚀性	corrosion resistance
耐脱漆剂性	paint stripper resistance
耐液性	fluid resistance
逆流漂洗	countercurrent rinsing
浓差极化	concentration polarization
pH 计	pH meter
pH 值	pH value
排电流保护	electrical drainage protection
排气洗净装置	exhaust air washing system
喷漆室	spray booth
喷砂	sand blasting
喷涂	spray coating
喷丸	shot blasting/peening

喷洗	spray rinsing
硼化	boriding
贫铬	chromium depletion
平衡电极电势	equilibrium electrode potential
起泡	blister
起皱	wrinkling
气候老化	weathering
气体腐蚀	gaseous corrosion
迁移数	transport number
铅笔硬度试验	pencil hardness test
浅坑	crater
桥式烘干室	bridge type baking oven
氢脆	hydrogen embrittlement
氢去极化	hydrogen depolarization
轻度缺陷	minor defect
清漆	varnish
去极化	depolarization
去离子水	deionized water
全浸试验	immersion test
热镀锌	galvanizing
热风干燥	hot air drying
热浸镀	hot dip

热浸镀锌层	hot – dip galvanized coating
热扩散	thermal diffusion
热喷涂	hot/thermal spraying
热熔	hot melting
人工干燥	artificial drying
人造海水	artificial sea water
韧性	toughness
溶解度	solubility
溶解氧	dissolved oxygen
乳化	emulsification
乳化除油	emulsion degreasing
润湿剂	wetting agent
散装涂镀	bulk coating
色差	colour difference
色牢度	colour fastness
闪镀/飞边	flash
十六醇	cetyl alcohol
试验板	test panel
室温固化	cold curing
刷镀	brush plating
刷涂	brush application/coating
双电层	electric double layer

水合封孔处理	hydro – thermal sealing
水洗	rinsing
水蒸气封孔处理	steam sealing
酸洗	pickling
体积电流密度	volume current density
添加剂	additive
铁锈	rust
通过式烘干室	passing through type baking oven
铜绿	patina
涂层附着力	adhesion of coating
涂层结合力	cohesion of coating
涂底漆	primer coating
涂覆/涂层	coating
涂覆工艺	coating process
涂料	paint
退镀	stripping
脱层/剥离	delamination
脱色	decolorization
外观	appearance
纹理方向	lay
物理气相沉积	physical vapor deposition (PVD)
雾化	atomizing

牺牲阳极	sacrificial anode
稀释剂	diluent
稀释稳定性	dilution stability
下吸式喷漆室	down draft spray booth
硝基漆	nitrocellulose lacquer
压送式喷枪	pressure feed type spray gun
严重缺陷	major defect
盐桥	salt bridge
盐雾试验	salt spray test
盐浴沉积	salt bath deposition
阳极	anode
阳极（氧）化	anodizing
阳极保护	anodic protection
阳极极化	anodic polarization
氧化	oxidation
氧化-还原电位	redox potential
氧化皮	scale
阴极	cathode
阴极保护	cathodic protection
阴极电泳涂装	cathodic electro deposition
阴极反应	cathodic reaction
阴极极化	cathodic polarization

应力腐蚀	stress corrosion
硬度	hardness
有机溶剂除油	solvent degreasing
预处理	pretreatment
预镀	preplating
原电池	galvanic cell
粘稠度	consistency
粘度	viscosity
遮蔽	masking
遮蔽胶带	masking tape
折叠	lap
针孔	pinhole
正极	positive electrode
致命缺陷	critical defect
中性盐雾试验	neutral salt spray (NSS) test
重力式喷枪	gravity spray gun
周期转向电镀	periodic reverse plating
贮存稳定性	storage stability
转化膜	conversion coating
着色	colouring
自动喷涂机	automatic spray machine
总酸度	total acid

质量管理
Quality Management

报废	scrap
不符合项报告	nonconformance report（NCR）
不合格	nonconformity
参与	involvement
测量	measurement
测量管理体系	measurement management system
产品技术状态信息	product configuration information
产品质量先期策划	advance product quality planning（APQP）
程序	procedure
持续改进	continual improvement
创新	innovation
等级	grade
法定要求	statutory requirement
法规要求	regulatory requirement
返工	rework
返修	repair
方针	policy
放行	release

风险	risk
改进	improvement
更改控制	change control
供方	provider/supplier
顾客	customer
顾客满意	customer satisfaction
观察员	observer
规范	specification
过程	process
过程控制	process control
过程质量控制	input process quality control (IPQC)
合格	conformity
合格产品目录	qualified products list (QPL)
合同	contract
活动	activity
积极参与	engagement
绩效	performance
计量确认	metrological confirmation
计量特性	metrological characteristic
计量职能	metrological function
记录	record
技术专家	technical expert

技术状态	configuration
技术状态管理	configuration management
技术状态管理机构	configuration authority
技术状态基线	configuration baseline
技术状态项	configuration object
监视	monitoring
检验	inspection
鉴定过程	qualification process
降级	degrade
经销商	distributor
纠正	correction
纠正措施	corrective action
可靠性	reliability
可信性	dependability
可追溯性	traceability
客观证据	objective evidence
库存控制系统	inventory control system
联合审核	joint audit
零缺陷	zero defect
每百万件中的不合格件数	non – conforming parts per million（ppm）
目标	objective

能力	capability
偏离许可	deviation permit
评审	review
全面质量管理	total quality control
缺陷	defect
确定	determination
确认	validation
让步	concession
人为因素	human factor
设计和开发	design and development
审核	audit
审核发现	audit finding
审核范围	audit scope
审核方案	audit programme
审核计划	audit plan
审核结论	audit conclusion
审核委托方	audit client
审核员	auditor
审核证据	audit evidence
审核准则	audit criteria
审核组	audit team
生产批	manufacturing lot

生产批号	manufacturing lot number
失效模式与影响分析	failure model and effect analysis（FMEA）
使命	mission
试验	test
首件鉴定	first article inspection（FAI）
首件确认	first article assurance（FAA）
受审核方	auditee
特性	characteristic
统计过程控制	statistical process control（SPC）
外部	outsource
物料需求计划	material requirement planning（MRP）
相关方	interested party/stakeholder
项目	project
项目管理	project management
项目管理计划	project management plan
效率	efficiency
协会	association
信息系统	information system
验收/合格质量标准	acceptable quality level（AQL）
验证	verification
要求	requirement
有效性	effectiveness

预防措施	preventive action
愿景	vision
战略	strategy
质量	quality
质量保证	quality assurance（QA）
质量保证条款	quality assurance provision
质量策划	quality planning
质量方针	quality policy
质量改进	quality improvement
质量功能展开	quality function deployment（QFD）
质量管理	quality management（QM）
质量管理系统	quality management system
质量计划	quality plan
质量检查	quality checking
质量控制	quality control（QC）
质量目标	quality objective
质量审核	quality audit
质量手册	quality manual
质量特性	quality characteristic
质量信息	quality information
质量要求	quality requirement
组织	organization

轴承
Bearing

安装槽	swaging groove
钡基脂	barium base grease
壁厚差	thickness variation
变形	deformation
表面缺陷	surface discontinuities
表面硬化轴承钢	case – hardened bearing steel
薄壁轴承	thin – walled bearing
超大型轴承	extra large size bearing
衬垫	liner
衬垫剥离强度	peeling strength of the liner
衬垫密合度	adhesion of the liner
衬垫粘接完善性	bond integrity of the liner
成对双联轴承	bearing in pairs
除锈	derusting
脆性断裂应力	brittle fracture stress
单列轴承	single row bearing
单一内径变动量 Δds	single plane mean bore diameter deviation

单一平面平均内径变动量 Δdmp	single plane mean bore diameter deviation
单一平面平均外径变动量 ΔDmp	single plane mean outside diameter deviation
单一外径变动量 ΔDs	deviation of a single outside diameter
电化学腐蚀	electrochemical corrosion
调心轴承	self – aligning bearing
定位槽/环槽	circumferential groove
动载荷	dynamic load
端面跳动	end face runout
多列轴承	multirow bearing
耳轴轴承	ball socket bearing
防尘盖	shield
防尘盖轴承	shielded bearing
防锈涂层	antirust coating
飞机操作系统轴承	bearing for aircraft control system
飞机机架轴承	airframe bearing
飞机起落架轴承	bearing for aircraft landing gear
分离式球轴承	separable ball bearing
腐蚀疲劳	corrosion fatigue
杆端关节轴承	rod end spherical plain bearing

杆端体	rod ends
钢对钢向心关节轴承	steel – steel radial plain spherical bearing
钢球	steel ball
高碳铬轴承钢	high carbon chromium bearing steel
公称直径	nominal diameter
公制轴承	metric bearing
沟	groove
沟道半径	groove radius
沟心距	axial distance between the two center lines of the innerring grooves
沟形误差	groove deviation
固体润滑剂	solid lubricant
关节轴承	spherical plain bearing
滚道	raceway
滚动体	rolling element
滚动轴承	rolling bearing
滚动轴承额定寿命	basic rating life of a rolling bearing
滚动轴承寿命	bearing life
滚针	needle roller
滚针轴承	needle roller bearing
滚子	roller
滚子轴承	roller bearing

过渡	transition
过渡配合	transition fit
过盈	interference
过盈配合	interference fit
航空发动机轴承	bearing for aircraft engine
滑动轴承	friction bearing
化学除锈	chemical derusting
化学腐蚀	chemical corrosion
划痕/刮痕	score/scratch
基本额定动载荷	basic dynamic load rating
基孔制	hole – basis system
基油	base oil
基轴制	shaft – basis system
基准孔	basic hole
基准轴	basic shaft
极限尺寸	limit size
间隙配合	clearance fit
键槽	longitudinal groove
角接触球面滚子轴承	angular contact spherical roller bearing
角接触球轴承	angular contact ball bearing
角接触推力关节轴承	angular contact thrust spherical plain bearing
角接触推力球轴承	angular contact thrust ball bearing

角接触轴承	angular bearing
接触角	contact angle
径向变形	radial deformation
径向额定静载荷	radial permissible static load
径向极限静载荷	radial ultimate static load
径向跳动	radial runout
径向游隙	radial clearance
静载荷	static load
聚四氟乙烯	polytetrafluoroethylene(PTFE)
可分离轴承	separable bearing
可靠性	reliability
宽内圈	extended inner ring
宽内圈轴承	bearing with extended inner ring
联合载荷	combined load
磷化处理	phosphate coating
螺纹直径	diameter of thread
螺旋桨轴承	airscrew bearing
满装球轴承	full type ball bearing
密封圈	seal
密封圈轴承	sealed bearing
密封性	sealing
磨损	wear/abrasion

磨损试验	abrasion test
内螺纹	internal thread
内圈	inner ring
内圈端面外径	outside diameter of inner ring face
内圈宽度	inner ring width
内圈内径	bore diameter of inner ring
耐腐蚀钢	corrosion – resistant steel
耐磨强度	abrasion strength
配合公差	fit tolerance
配合极限	fitting limit
喷砂清理	sand – blasting
疲劳剥落	fatigue flaking
疲劳载荷	fatigue loading
偏心角	eccentric angle
偏转角度	angle of tilt
偏转力矩	swivelling/tilting torque
使用寿命	service/working life
平均寿命	average life
轻系列轴承	extra light series bearing
球	ball
球面滚道	spherical raceway
球面滚子轴承	spherical roller bearing

球形偏差	deviation from spherical form
球直径	sphere diameter
球轴承	ball bearing
曲率半径	curvature radius
润滑槽	lubrication groove
润滑孔	lubrication hole
润滑油	lubricating oil
润滑脂	grease
三点接触球轴承	three point contact ball bearing
沙尘试验	sand and dust test
深沟球轴承	deep groove ball bearing
渗碳层	carbonization zone
渗碳轴承钢	carburized bearing steel
室温运转状态	Behaviour in rotation at ambient temperature
双列调心球轴承	double row self – aligning ball bearing
双列角接触球轴承	double row angular contact ball bearing
双列向心球轴承	double row radial ball bearing
双列圆锥滚子轴承	double row tapered roller bearing
双列轴承	double row bearing
双向推力球轴承	double – direction thrust ball bearing
四点接触球轴承	four point contact ball bearing
四列圆锥滚子轴承	four row tapered roller bearing

酸洗	pickling
弹簧圈/卡环	spring ring
弹性变形	elastic deformation
碳氮共渗	carbonitriding
套圈	bearing ring
同轴度	axiality
凸缘轴承	flanged bearing
推力滚针轴承	needle roller thrust bearing
推力球轴承	thust ball bearing
推力圆柱滚子轴承	cylindrical thrust roller bearing
推力轴承	thust bearing
椭圆轴承	elliptical bearing
外螺纹	outside thread
外球面	spherical outside surface
外圈	outer ring
外圈宽度	outer ring width
外圈外径	outside diameter of bearing outer ring
外圈外圆柱表面	cylindrical outside surface of an outer ring
微型轴承	miniature bearing
稳定性	Retention
无载起动力矩	starting torque without load
向心关节轴承	radial spherical plain bearing

向心球轴承	radial ball bearing
向心轴承	radial bearing
小型轴承	extra small bearing
形状公差	form tolerance
形状误差	form deviation
旋转精度	running accuracy
压碎载荷试验	crushing load test
盐雾试验	salt spray test
英制轴承	inch bearing
永久变形	permanent deformation
油封	preservation
圆柱度	cylindricity
圆柱滚子	cylindrical roller
圆柱滚子轴承	cylindrical roller bearing
圆柱孔轴承	bearing with cylindrical bore
圆锥孔轴承	bearing with tapered bore
允许磨损	allowable wear
支承应力	bearing stress
直径	diameter
止动环	snap ring
中大型轴承	medium size bearing
中小型轴承	small size bearing

轴	shaft
轴承	bearing
轴承安装	bearing installing
轴承保持架	bearing cage
轴承拆卸	bearing removal
轴承衬套	bearing lining/liner
轴承承载能力	bearing power
轴承单元	bearing unit
轴承钢	bearing steel
轴承滚道表面剥落	bearing flaking
轴承过载	bearing over loading
轴承疲劳极限	bearing fatigue point
轴承锁紧垫圈	bearing lock washer
轴承锁紧螺母	bearing locknut
轴承凸缘	bearing flange
轴承维护	bearing maintenance
轴承性能	bearing performance
轴承咬死	bearing seizure
轴承游隙	bearing clearance/play
轴承座	housing
轴肩	axle shoulder
轴颈	axle neck